MW00427036

Essentials of the Orthopaedic Examination

Essentials of the Orthopaedic Examination

Alan Hammer FRCS (Ed), FCS (SA), FCS (SA) (Orth)

Consultant Orthopaedic Surgeon, Addington Hospital,
Durban, South Africa

Edward Arnold
A member of the Hodder Headline Group
LONDON BOSTON MELBOURNE AUCKLAND

First published in Great Britain in 1995 by
Edward Arnold, a division of Hodder Headline PLC,
338 Euston Road, London NW1 3BH

Distributed in the USA by Little, Brown and Company,
34 Beacon Street, Boston, MA 02108

Whilst the advice and information in this book is believed to be true and
accurate at the date of going to press, neither the author nor the publisher
can accept any legal responsibility or liability for any errors or omissions
that may be made. In particular (but without limiting the generality of the
preceding disclaimer) every effort has been made to check drug dosages;
however, it is still possible that errors have been missed. Furthermore,
dosage schedules are constantly being revised and new side effects
recognised. For these reasons the reader is strongly urged to consult
drug companies' printed instructions before administering any of the drugs
recommended in this book.

British Library Cataloguing in Publication Data
A catalogue record for this book is available from the British Library

ISBN 0 340 61382 3

1 2 3 4 5   95 96 97 98 99

Typeset in Times by GreenGate Publishing Services, Tonbridge, Kent
Printed and bound in Great Britain by J.W. Arrowsmith Ltd., Bristol

Contents

Preface   vi

Introduction   vii

1   Orthopaedic history   1

2   General examination   13

3   Examination of the spine   26

4   Examination of the upper limb   40

5   Examination of the lower limb   64

6   Laboratory investigations   92

7   Imaging investigations   103

8   Orthopaedic trauma   108

Recommended reading   112

Index   113

Preface

Orthopaedic surgery is an ever-enlarging clinical discipline, but the time allocated to the study of this subject in the average medical school is relatively low. It is thus difficult for the student to learn as much of this subject as would be desirable.

Much can be learnt directly from the patient. The aim of this book is to direct the student in the clinical situation in the hope that it will assist him, or her, to gain as much insight into the subject as possible in the time available.

An attempt has been made to keep the scheme simple. In this respect thanks must go to Mr Graham Apley who propounded the system of 'Look, Feel and Move'. In this book a further parameter, 'Measure' has been added.

Introduction

'It is much more needful to teach people the art of seeing.'

Carl Gustav Jung

An accurate diagnosis is the first step in the successful treatment of any disease condition. This applies to orthopaedic surgery as much as to any other medical discipline. To arrive at a diagnosis, information gleaned from a relevant history and examination is essential.

For any area in the body, be it a joint, a muscle, a tendon, or a nerve, there is a limited number of different pathologies that may affect it. The enquiry, once the area of concern has been delineated, should be directed at determining which possible pathological processes is actually present. It is thus necessary to have a working knowledge of the possible pathologies that may be present before undertaking the physical examination.

A useful way of directing the history and examination is to categorise any problem into one of the following pathological groups:

- congenital
- infective
- vascular
- auto-immune
- neurological.
- traumatic
- neoplastic
- metabolic
- degenerative

The history

The 'history' is the speaking part of the assessment and has two aims. The first is to determine the course of events preceding the examination and, by considering what has happened, to make some sort of assessment of the cause of the problem. The second aim is to gain insight into the attitude and health of the patient in order to plan and institute his or her treatment.

The situation and circumstances of individuals vary considerably. The treatment ultimately selected may or may not be the best available at that time, but it should be the most appropriate for that particular individual.

The physical examination

The aim of the physical examination is to find the physical signs present that will confirm or modify the initial diagnosis suggested by the history.

What is found should correlate with what has been said. If there is a discrepancy between the history and physical examination someone is wrong. It will be either the patient who is wrong, or it will be yourself who is wrong in your assessment.

A distinction needs to be made between 'looking' and 'seeing'. Anyone can 'look' at a physical sign. It is only the person who knows what they are looking at and who understands its relevance who will 'see' the physical sign in its context.

The assessment

The assessment combines what is determined from the history with what is elicited during the physical examination in order to arrive at a conclusion as to what is most likely to be wrong with the patient – the diagnosis. It is only when one has an idea of the nature of the underlying pathology that effective therapy for the problem can be instituted.

To interpret the facts, a solid background of basic knowledge in anatomy, physiology and pathology is essential. The importance of these subjects cannot be overemphasised. It is necessary, many times, to refer back to them to determine the relevance of a particular sign. As these topics tend to be taught as academic subjects during the preclinical years, the student has to relearn them, with a clinical bias, during the clinical years.

It is often only with experience that one comes to learn to interpret the signs and symptoms presented by the patient and fully understand their meaning and significance.

1 Orthopaedic history

(The patient's **subjective** experience)

The medical history and examination is a confidential and intimate communication between doctor and patient, so it should be conducted in an area that offers the necessary privacy.

The 'history' defines the nature of the complaint, as it is experienced by the patient. It is obtained by allowing the patient to describe what has happened. (Once a patient is talking, there is a tendency for them to digress and it is often necessary to guide the conversation by using more or less direct questioning.)

The 'history' allows you to determine what, and over what period, has happened to the patient. A number of important questions need to be answered.

Social and epidemiological

What is the patient's name?   It is important to approach the patient as an individual with normal hopes, aspirations, fears and doubts. It is highly probable that the patient is acutely disturbed by the process taking place in his body.

What is the patient's age?   Orthopaedic problems present as congenital lesions in the very young or as degenerative lesions in the middle aged and elderly. Lesions due to infection or tumour occur over a wide age range but are often age related. In many cases these lesions, such as some tumours, are very specific for the age of the patient. Injuries can occur at any age although the more severe (high velocity) injuries are more common among the young and the less severe (low velocity) injuries are more common among the elderly.

What is the patient's sex?   Some conditions are sex related. Typical of these are the inherited sex-linked conditions carried by the mother and manifest in the son. Similarly, fractures due to osteoporosis occur more commonly in women than in men.

Where does he or she live?   This is a good indicator of the economic status of the patient and can imply a way of life that may be relevant to the present complaint. For instance, infective processes, acute and chronic, have long been associated with conditions of social deprivation or personal neglect.

Is there a possibility of exposure to infective or parasitic agents? There may be a similar disease in a family member or the individual may live in an area where specific diseases are endemic.

Presenting complaints

The patient usually presents with one or more of the following complaints:
- pain
- deformity
- loss of function

Pain

Pain is not normally experienced in everyday life, but it is often associated with various pathological conditions. It is often the pain that first brings the individual's attention to the presence of a pathological condition in his body.

The pain sensation varies with the type of pathological process. The character of the pain may reflect the nature of the pathology which is present. The acute pain of movement of an injury or arthritic joint differs from the constant, sharp, throbbing pain of an abcess, which differs from the constant, dull, nagging pain associated with an underlying malignancy.

Questions that may help differentiate the pain and give an idea of the underlying process are as follows:

Where is the pain?   Having an idea of the anatomical structure in which the pain may exist is a good start. Pain in a joint may have a different cause to pain felt in a bone or in muscle.

When did it start?   The period the pain has been present is important and the patient may relate its onset to a specific event. One must be careful in accepting a condition in relation to a specific event. In some cases it is relevant, such as conditions arising from injury. However, in some cases it is the attempt by the patient to rationalize the problem, and the episode has no bearing at all and may serve only to mislead.

How did it start?   There is often a big difference between a condition arising acutely from a specific event (such as an accident or vascular occlusion), and one that develops insidiously (such as a malignancy or inflammatory condition).

How severe is it?   The pain resulting from a sprain or fracture may be incapacitating. However, it is suprising how much pain some people can tolerate in association with a degenerative condition. Perhaps this is because the pain develops slowly and the patient adapts to it. The degree of pain often influences the choice of treatment – the greater the pain the more aggressive the therapy may need be.

What is its nature?   In orthopaedic surgery two main types of pain are experienced – a sharp, acute pain (usually from injury) and a constant, dull, nagging pain (usually from an inflammatory process or tumour).

Is it constant or intermittent?   Pain due to an infection or malignant tumour, is generally constant. An abcess may give a throbbing pain.

Does it radiate?   Radiation of pain down a limb is typical of nerve root entrapment (radicular pain). Radiation down the leg is termed 'sciatica'. Referred pain arises in a deep structure and is felt by the patient to be in another area.

It is sometimes difficult to differentiate radicular from referred pain and this has to be differentiated by the physical examination. Sensory and motor loss, in the distribution of the affected nerve, will be found in association with radicular pain.

What aggravates the pain?   Pain that arises from a sprain of a joint or fracture of a bone is typically aggravated by movement of that structure. (In the back this pain is termed 'mechanical'.) Similarly, degenerative osteoarthritis may be aggravated by movement. The pain resulting from an infection or a tumour is often not affected by movement.

What relieves the pain?   The answer may give an indication as to previous therapy for the condition and to its effectiveness. Relieving factors may take the form of drugs (such as analgesics or antibiotics) or of mechanical factors, such as splints or casts which rest and support the affected area. Avenues for possible lines of future treatment may be indicated by these measures.

Is it present at night and/or does it disturb sleep?   Painful musculo-skeletal structures are often splinted by involuntary muscle spasm which reduces the pain. However, during rest or sleep this spasm tends to subside, allowing the affected structure to move and initiating or aggravating the pain.

Deformity

The complaint of deformity is commonly obtained from the parents of a child who are concerned with an apparent deviation from normal. The child himself adapts to a deformity and often there is little functional defect. In the older person the deformities are usually recently acquired and may be associated with complaints of pain or loss of function.

Questions related to deformity are as follows:

Where is the deformity?   Deformities of the spine, upper limb and lower limb have specific characteristics with regard to appearance and function, or loss thereof.

What is its nature?   Angular, rotatory, shortening, lumps, or gaps. The patient is usually concerned only that there is an angulation, twisting or protuberance in the affected region. The nature of the deformity is related directly to the underlying pathology and, in many cases, indicates the cause for the condition: whether congenital, traumatic, inflammatory or neoplastic.

How long has it been present?   The difference between congenital and acquired deformities is profound and may greatly influence what can be done to correct the condition. Congenital deformities are often associated with absence or maldevelopment of tissues. Acquired deformities have occurred in tissues that were initially normal.

How did it start?   A sudden, traumatic event may leave the patient with a deformed limb or spine. Gradual deformities are often associated with continuing conditions affecting the growth of the region.

Has it been progressive and, if so, how rapidly?   Causes of progressive deformity include those due to a neurological lesion, giving unbalanced muscular action; those from abnormalities of the growth plate, giving unbalanced growth; those due to softening of bones, causing them to bend; and those due to progressive scarring of the area, causing contractures. Ask questions about neurological conditions and injuries, particularly those to the head and the limb.

Joint deformities are often due to muscle imbalance resulting from a neurological conditions such as poliomyelitis or cerebral palsy in children. In adults, cerebrovascular episodes often lead to neurological imbalance. Muscle imbalance over a joint leads inevitably to a crippling deformity as the joint becomes contracted on one side. In these cases it is necessary to differentiate muscle paralysis from muscle spasticity, although both causes can lead to imbalance if only one muscle group is affected. In cases of muscle spasticity the spastic muscle overwhelms its normal antagonists, which are effectively paralysed. (A joint where both protagonist and antagonist muscles are paralysed does not become deformed but remains flail.)

Partial injury of the growth plate (physis) will cause the bone to grow towards the side of injury. Complete loss or dysplasia of the growth plate will cause stunting of that limb, with or without an angular deformity.

Softening of the bones, particularly those in the lower limbs, leads to a progressive bowing of the limbs as gravity and muscle contractions act on them.

Injury of the soft tissues about a joint can lead to fibrosis and contraction of the skin, muscles or tendons. These are seen in contractures due to burns, and in scarring after injuries and infections affecting muscles and tendons. A particularly pernicious cause for progressive contracture of a joint is a Volkmann's ischaemic contracture following ischaemia to the muscles.

Are any associated conditions still present?   A progressive deformity may indicate that the causative process is still present and active, and that treatment of this may be necessary.

Paget's disease and osteogenesis imperfecta resulting in bone softening are difficult to treat. However ricketts may respond to medical therapy.

Bone dysplasias, osteoporosis and tumour deposits weaken the bone, leading to fractures of the vertebrae and the condyles of joints. In these cases there will be progressive deformities of the joints or the spine.

Rheumatoid arthritis, infections and other conditions that destroy the cartilage lead to a progressive deformity of the affected joint.

N.B. Growth plate and other deformities associated with growth will progress only as long as growth is occurring in that region.

Loss of function

Orthopaedics is intimately concerned with the function of the body. A guiding principle of the AO/ASIF (Association for the Study of Internal Fixation) is the aphorism:

'Life is Movement, Movement is Life'.

The loss of function experienced by the patient will depend on the nature and area of the body affected by the pathological condition. Loss of function is variable and relative to the individual patient. Some can adapt to the most major orthopaedic problem, while others are incapacitated by a seemingly minor condition. In many cases loss of function is incidental to the main presenting complaint of pain or deformity.

Loss of function of the arms will affect the individual's ability to grasp and manipulate objects. Loss of function of the legs will impair walking, running, climbing, etc. Loss of function from a spinal lesion may affect either the arms, or legs, or both.

Questions regarding loss of function are similar to those related to deformity:

What function is lost?   The specific functional loss may indicate the anatomical area affected. It is important to differentiate between functional loss caused by pain, deformity or paralysis. Question specifically the function of the affected area.

Questions related to the function of the hand pertain to its two main functions – pinch and grasp. Can the patient hold a pen, or cup? Can he hold and lift a parcel?

Questions related to the shoulder and elbow relate to the ability to position the hand for use. Can the patient comb his hair? Can she do up her brassiere?

Questions on the functional assessment of the hips and knees relate to locomotion. How far can the patient walk? Can he climb stairs? Can she reach her foot?

What is its effect?   The loss of function experienced by the patient may affect social, occupational or leisure activities and this may be specific for the affected region.

How did it start?   The loss of function may be associated with a specific event (such as an accident) or may have developed slowly and insidiously with no obvious associated factor (as occurs in a degenerative arthritis).

When did it start?   Sudden loss of function can be calamitous to the patient. Long standing loss will have allowed the patient to adapt to the condition, both physically and psychologically.

Has it been progressive?   As with deformity, neurological pathology, as occurs with injury, infection and tumour of the nervous system, is a potent cause for progressive loss of function. When there is muscle imbalance the joint then becomes contracted and functionless. Similarly, degenerative conditions of joints or muscles lead to an insidious loss of function.

Are aids to function used?   The use of calipers, walking sticks, crutches and other supports indicates the forms of previous therapy introduced for the condition and the success, or lack thereof, of the adaptive changes that may have been made by the patient.

Chronic affliction of joints

When long standing pathology is present in a joint, specific information must be obtained:

Does the joint swell and, if so, how often?   Various chronic inflammatory and degenerative conditions cause the secretion of excessive quantities of fluid within a the joint, or of the accumulation of pus within its cavity. This fluid distends the joint, leading to an obvious swelling.

Does the joint 'give way'?   Laxity of the constraining ligaments of a joint, from trauma or erosion, will allow the joint surfaces to slide under stress. The proprioceptive nerves detect this movement. When this occurs under load the patient feels as if the joint is 'giving way'.

Does the joint 'lock'?   Various abnormal tissue fragments may find their way between the articulating surfaces and cause a mechanical obstruction to the movement of the joint. These include free floating cartilagenous bodies, osteochondral fragments, or damaged portions of ligaments or menisci. The result is a joint that 'locks' intermittently at some part of its range of motion.

General orthopaedic history

Following on from factors related to the localised orthopaedic problem, there may be generalised problems that are directly or indirectly related to the presenting complaint.

Generalised orthopaedic problems Determine whether other joints or bones are affected. A condition that affects a single joint (such as a septic arthritis) is very different from a condition that can affect many joints (such as rheumatoid arthritis). A generalised condition may indicate a generalised congenital mesenchymal defect (such as osteogenesis imperfecta or a bone dysplasia).

Occupation

What is the occupation of the patient? Orthopaedic surgery is primarily concerned with function, and the work of the patient is of major importance.

Among those who are employed, impairment of the ability to work (from pain, deformity or paralysis) is often the most disturbing factor for the patient. This is not such a pressing problem amongst those who are retired. It is also necessary to determine whether the work itself has been responsible for the condition affecting the patient.

Past and present surgical history

What orthopaedic surgical procedures have been performed for the present condition? If specific for the presenting complaint these procedures will indicate what treatments have already been attempted for the problem. Operations of this nature may include:

- Osteotomies (bone or joint diseases)
- Arthroplasties (generalised joint disease)
- Fracture fixations (trauma or generalised bone disease)
- Muscle or tendon transfers (polio or motor neurone disease).

What other surgical procedures have been performed? These may have relevance to the orthopaedic condition, i.e.:

- *General surgical procedures* (e.g. bowel resections) These may suggest conditions such as Crohn's disease or ulcerative colitis which have joint manifestations. Similarly, a previous gastrectomy may lead to malabsorption, especially that of calcium, leading to osteoporosis.
- *Gynaecological procedures* Oestrogen deficiency is linked to impaired bone turnover. A procedure such as hysterectomy may thus be related to a generalised osteopaenia.

Past and present medical history

Past or present medical conditions may have relevance to the musculo-skeletal presentation. Ask questions in the following areas:

Fevers These are characteristic of many infective or inflammatory processes that directly or indirectly affect the musculo-skeletal system.

Skin rashes These are found often with the collagenoses. Psoriasis shows as patchy, flaking, erythematous areas of skin and is commonly associated with large joint arthritis. Lupus erythematosis produces a classical deep red skin on the face or hands. Transient erythematous rashes may be associated with Still's disease and rheumatic fever. Skin rashes may also complicate the administration of non-steroidal, anti-inflammatory drugs.

Respiratory system Question particularly shortness of breath. The patient may be suffering from obstructive airways disease which, even if not related to the orthopaedic problem, will increase the problems of operative management.

Cardiovascular system Symptoms of recent or previous chest pain, palpitations, dyspnoea at rest or with effort suggest cardiac pathology. A history of previous deep vein thrombosis places the patient at a high risk of recurrence following surgery.

Diabetes The presence of diabetes often complicates any orthopaedic condition. Diabetic neuropathy leads to loss of sensory and motor functions. The metabolic effects of this condition are such that tissue repair is frequently compromised and infection is an ever present threat, as the tissue response to bacterial invasion is poor. (Note that oral antidiabetic agents can affect the blood pressure of the patient during anaesthesia and the anaesthetist should know that the patient is taking them.)

Gout This is not necessarily confined to the big toe but may affect any joint of the body. The condition leads to pain and swelling of a joint, and a gouty process can mimic a septic arthritis. Often, in severe gout, swellings and deposits of uric acid are seen under the skin around joints, or over the ears and behind the elbows. (A history of previous diagnosis of gout is important as many patients discontinue their uricosuric agents once they feel better and often do not relate their present problem to gout as it may take place many years later.)

Collagen disorders (e.g. rheumatoid arthritis) These are generalised conditions affecting all the mesenchymal tissues of the body, not only the joints. Not only may the deformities and disabilities in the large joints be evident but involvement of the smaller joints can be very disabling. Involvement of the upper limbs may impair use of crutches. Involvement of the temporomandibular joints and those of the neck may make intubation during anaesthesia hazardous. Many patients have been receiving steroid therapy.

This in itself impairs wound healing and wound breakdown, and infection is common in these patients.

Gastro-intestinal problems Determine that the patient is eating normally and that there has been no loss of weight. Obesity makes surgery technically difficult and increases the risk of surgery for the patient. The collagenoses, particularly ulcerative colitis and Reiter's syndrome, may present with a diarrhoea.

Urinary symptoms These are often associated with musculo-skeletal conditions. Recent haematuria may indicate an urinary tract infection or a renal tumour presenting as backache. Renal calculi occur in gout and hyperparathyroidism. Septic arthritis complicates gonococcal infection. Reiter's disease, with urethral discharge, may present with osteoarthritis of a large joint. Acute urinary retention can be associated with a central prolapsing intervertebral disc. Paraplegic patients risk recurrent urinary tract infections. (Note – Urinary tract infection is a contra-indication to joint replacement surgery.)

Bleeding tendency The presence of a bleeding diathesis can present as pain, swellings and bruisings in the soft tissues and joints. Chronic bleeds into a particular joint lead to its degeneration and arthritic changes. (Note – It is disastrous to discover the patient's bleeding or clotting problem in the middle of surgery.)

Neurological Previous strokes or injuries to the head or peripheral nerves may be relevant to the present orthopaedic problem. Parkinsonism affects the musculo-skeletal system as a whole.

Family history

Has the patient a history of unusual orthopaedic problems or deformities in near relatives? – e.g. hypophosphataemia. Patients may exhibit a genetic problem in varying degrees of penetrance. The orthopaedic presentation results from effects of these inherited conditions on the mesenchymal tissues (connective tissue, muscle and bone). Note that some conditions arise by spontaneous mutation. Examples of this can be found in haemophilia, achondroplasia and osteogenesis imperfecta.

Drug history

The present and past drug history of the patient can give insight into the duration and course of the condition, and give some idea as to the effect of previous therapy. Enquire particularly into the prolonged use of steroids, analgesics and alcohol.

Chronic steroid ingestion generally affects the structure of mesenchymal tissue, leading to osteoporosis, poor wound healing and decrease resistance to infection. It can also initiate avascular necrosis of the femoral head.

Chronic analgesic ingestion   can impair renal function and can lead to an anaemia as a result of chronic blood loss from gastric erosions.

Chronic alcohol ingestion   affects orthopaedic management in a number of ways. Alcohol intake is often associated with violence and injury. Chronic alcohol ingestion affects the metabolism of bone both directly and indirectly, leading to a severe osteoporosis. Liver function may be impaired, leading to defects of coagulation which can affect surgery. Postoperative management of the patient can be complicated by the patient developing delirium tremens. Similarly alcoholism can lead to avascular necrosis of the femoral head.

Degrees of pain

Pain may be graded into five grades, depending on the quantity of analgesics taken by the patient:

Grade 0   no pain
Grade 1   minimal, occasional pain requiring no analgesics
Grade 2   mild pain controlled by occasional use of analgesics
Grade 3   more severe pain, controlled by regular ingestion of analgesics
Grade 4   constant, severe pain, requiring high levels of NSAIDs or opiates for control

---

Orthopaedic history summary

Social and epidemiological

What is the patient's name?
What is the patient's age?
What is the patient's sex?
Where does he or she live?
Is there a possibility of exposure to infective or parasitic agents?

Presenting complaints

The patient presents with one or more of the following complaints:

    pain
    deformity
    loss of function

Pain

Where is the pain?
When did it start?
How did it start?
What is its nature?
How severe is it?
Does it radiate?
Is it constant or intermittent?
What aggravates the pain?
What relieves the pain?
Is it present at night?

Deformity

Where is the deformity?
What is its nature?
How long has it been present?
How did it start?
Has it been progressive?
Associated conditions.

Loss of function

What function is lost?
What is its effect?
How did it start?
When did it start?
Has it been progressive?
Are aids to function used?

Chronic affliction of joints

Does the joint swell?
Does the joint 'lock'?
Does the joint 'give way'?

General orthopaedic history

Is there any general musculo-skeletal condition present?

Occupation

What is the occupation of the patient?
What are his leisure activities?

Past and present surgical history

What surgical procedures are or have been performed, including:
osteotomies
arthroplasties
muscle or tendon transfers
fracture fixations
general surgical operations
gynaecological operations.

Past and present medical history

Fevers
Skin rashes
Diabetes
Gout
Collagen disorders
Gastro-intestinal problems
Urinary symptoms
Bleeding tendency
Respiratory system
Cardiovascular system
Neurological

Family history

Are there any unusual orthopaedic problems or deformities in near relatives?

Drug history

Steroid ingestion
Analgesic ingestion
Alcohol ingestion

2 Orthopaedic examination

(Your **objective** findings)

The examination is a confidential procedure during which it is necessary to have complete trust between the patient and yourself. Conduct the examination in a secluded area, reserved for this purpose. Try to ensure that the patient is as relaxed as possible.

Ask the patient to remove as much clothing as is necessary for you to see the affected area of the body. Screen off the patient if necessary and position him (or her) on a comfortable examination couch in an adequately warmed room. Ensure that the sheets are clean and that your hands and instruments are warm, before touching the skin.

Conduct the physical examination in three parts. First examine the patient generally over the whole body, noting the presence of any generalised disease or condition that may be relevant to the musculo-skeletal problem. Look then over the whole of the affected limb, or spine, in order to assess it in relation to the rest of the body. Finally, examine the local area of the presenting complaint; note your findings and assess them in relation to the whole of the limb and the rest of the body.

General systematic examination

Examine the whole patient. Note pathological conditions in different organs and systems that may affect, or be associated with, problems presenting in the musculo-skeletal system.

Note the following features:

State of health   Assess the health of your patient. The general appearance of the patient often conveys a sense of health or of non-health. His or her general state of nutrition, the colour of the skin, the way he or she moves, the air of confidence or defeat that the patient exhibits all convey indirect signs of health or otherwise.

In severe, acute infective processes, such as an osteomyelitis, peritonitis, or bronchopneumonia the patient appears decidedly ill. (Even in these days of powerful antibiotics a bacterial septicaemia is still a threat to the life of the patient.) Severe wasting of the body tissues may be seen in advanced illness, particularly when malignant disease is present.

State of nutrition   Note whether your patient is overweight (obese), thin and wasted (cachexic) or 'normal'. Problems can be anticipated in cases where there is over or under nutrition.

Obese patients make any surgical procedure a taxing technical exercise, as the surgical field is hidden by a large bulk of tissue. They are at risk for a venous thrombosis of the leg and also have a tendency towards infection in their operation wounds. Massive obesity, by the weight of the chest, impairs pulmonary and cardiovascular function during anaesthesia and puts the patient at major anaesthetic risk.

Diminished subcutaneous fat, muscle wasting and associated loss of body weight may be the consequence of inadequate food intake and may be seen with poverty, deprivation, alcoholism or drug dependency. A severe form of malnutrition seen occasionally among children in the tropics is kwashiorkor – a deficiency of protein accompanied by varying degrees of vitamin deficiencies. The child is ill, stunted and presents with hair changes; scaly, depigmented skin; cutaneous sores; and oedema.

The avitaminoses have their own characteristic signs, depending on the vitamin which is lacking.

- *Beriberi* (thiamine, vitamin B1 deficiency) presents predominantly with a peripheral neuritis (dry form), or with a cardiac myopathy and generalised fluid retention and oedema (wet form).
- *Ariboflavinosis* (riboflavin, vitamin B2 deficiency) there is angular stomatitis of the mouth, glossitis of the tongue, and seborrhoeic excrescences of the face.
- *Pellagra* (nicotinic acid deficiency) is associated with great emaciation. It presents mainly with intestinal atrophy (stomatitis and various pathologies in the intestine); cutaneous rash (a rough, hyperkeratotic skin with a patchy erythematous rash and petechiae); and central, demylinating neuritis. The condition is thus characterised by the triad – diarrhoea, dermatitis and dementia.
- *Scurvy* (vitamin C deficiency) is rarely seen in its florid state, although a subclinical state may be present. Its onset is insidious, with loss of weight, weakness and pallor, and stiffness in the leg muscles. The gums are affected and the teeth fall out. Subperiosteal and intramuscular haemorrhages occur.

Although it is often easier to expose the operative field in thin, malnourished individuals, tissue healing and resistance to infection are markedly reduced in these people.

State of hydration   Assess the state of hydration of your patient. Dehydration results from diminished fluid intake or excessive fluid loss and its cause must be elucidated. Although he may not volunteer it, a dehydrated patient is thirsty and will soon acknowledge this, if asked.

Skin turgor is reduced with dehydration and can be assessed by plucking up a fold of skin over the upper chest. Slow descent of the fold suggests dehydration but this test should be interpreted with caution in the elderly, in whom the subcutaneous tissues are atrophic. In severe cases the eyes are 'sunken'. The best clinical indicator of dehydration is the tongue. This looks and feels dry in the dehydrated patient.

Fluid retention may be the consequence of cardiac or renal disease. Generalised pitting oedema requires these organs be assessed. Localised oedema in a single limb may suggest venous thrombosis or a lymphatic obstruction. The Klippel Trelanney syndrome may manifest as generalised oedema or vascular malformations of a limb.

Generalised skin colour changes, blemishes, or rashes   Assess the patient for generalised changes in the skin. These may be associated with disease processes which are present incidentally.

Jaundice, if present, suggests hepatocellular dysfunction and the liver and spleen should be examined. Cyanosis of the legs suggests severe peripheral vascular disease requiring assessment of the pulses. (Peripheral vascular disease presents a risk of coronary thrombosis or a cerebrovascular accident during surgery and this fact may influence one's management of the patient.)

Other conditions have musculo-skeletal manifestations as well as cutaneous signs; for instance Psoriasis often has joint manifestations, while other 'collagen' diseases can present with a rash.

Respiratory system   Watch the way your patient is breathing. Tachypnoea may indicate anxiety, hypovolaemic shock, shock lung and the adult respiratory distress syndrome, all of which may follow injury or pulmonary embolus. Orthopnoea suggests left sided cardiac failure, as may occur in rheumatic fever. Nasal flaring and the use of accessory respiratory muscles indicates severe respiratory distress.

Chronic obstructive airways disease is a common problem, particularly in the elderly patient, and may add to the complications of anaesthesia if surgical management is planned.

Cardiovascular system   Measure the pulse rate. Tachycardia in a traumatised patient suggests hypovolaemia requiring urgent attention. Similarly a raised pulse rate may also be associated with the fever of a localised or generalised infection, or with cardiac disease.

Measure the blood pressure. Hypotension may be associated with cardiac infarction or hypovolaemia from major blood loss. (Do not forget that the heart or great vessels may be injured in people suffering severe trauma.) Hypertension suggests risk of a cerebrovascular stroke or cardiac thrombosis and thus an increased operative risk. Note that congenital deformities of the heart may be present in association with skeletal malformations.

Alimentary system   Inspect and palpate the abdomen and assess the stool. Acute gastric erosions, indicated by epigastric tenderness, haematemesis and malaena may result from chronic analgesic intake but concomitant chronic peptic ulceration must be excluded. The abdominal viscera may be injured in cases of polytrauma and particularly with fractures of the pelvis. As with the cardiovascular system, congenital malformations of the alimentary system may be associated with skeletal malformations.

Genito-urinary system   Palpate the abdomen and loins and assess the urine. The genito-urinary system can be traumatised, along with the musculo-skeletal system during injury. This is indicated by haematuria immediately after the accident.

Renal colic can occur directly as a result of two 'orthopaedic' conditions. Raised levels of urate secreted into the urine, in sufferers from gout, may precipitate out as symptomatic calculi. Similarly calcium phosphate calculi can occur in patients suffering from hyperparathyroidism or those who, from paralysis or weakness, are forced to prolonged recumbency.

Renal lesions, infection or tumours, may present as loin pain and resemble backache. Alternatively, malignant tumours of the genito-urinary system, kidney and prostate may metastasise and present with bone lesions. Again congenital malformations of the genito-urinary system may be associated with skeletal malformations.

Neurological   Assess the neurological status of the patient. Attention should be paid to pathology in the nervous systems of the patient: those affecting the central nervous system and those affecting the peripheral nervous system.

Parkinsonism, a neurological condition affecting one in 100 elderly patients, gives rise to a spastic dystonia of the muscles which can be incapacitating to the afflicted individual. Spasticity may also be associated with cerebral palsy in the young and a previous cerebro-vascular accident in the elderly.

Mental confusion can be associated with a cerebro-vascular episode, hypoxia, diabetes, renal or hepatic dysfunction, or senility. Delerium may be associated with fever or chronic alcohol ingestion.

Localised or generalised muscular paralysis may be associated with poliomyelitis or motor neurone disease. (The muscular dystrophies and myasthenia gravis may be considered here as part of the differential diagnosis.)

Injury, infections, tumours and degenerative conditions of the nervous system give rise to a variety of neurological signs and symptoms – motor and sensory – which may impair musculo-skeletal function.

General orthopaedic examination

Examine the whole patient. Note pathological conditions in the musculo-skeletal system that may affect, or be associated with, the specific problems presenting in the musculo-skeletal system.

Assess and record the following:

The attitude of the patient to his problem   Some people are positive about their affliction and it is usually fairly easy to involve them in their treatment. Others have difficulty in accepting the problem and can prove refractory to any treatment. (Note those who tend to hide the affected area under their clothes or bedding, or resist being examined.)

Size of the patient Body size varies tremendously and it is sometimes difficult to decide whether an individual is abnormal in stature. In these cases of doubt, tables and graphs are available to indicate the normal ranges of body size for the patient's age.

Excess secretion of growth hormone results in gigantism before epiphyseal closure and acromegaly after epiphyseal closure.

Diminished stature results when there is impairment to the normal growth of the body, or part of it. Although hormonal deficiencies (pituitary or thyroid) may result in dwarfism, more commonly there is a genetic defect affecting specific parts of the skeleton. This defect may be inherited or arise spontaneously as a mutation. The defect can affect mesenchymal tissues as a whole (osteogenesis imperfecta and mucopolysaccharidoses), or may manifest only in the bones.

The long bones may be affected in their epiphyseal region (epiphyseal dysplasia), their metaphyseal region (metaphyseal dysplasia), or in their diaphyseal region (diaphyseal dysplasia). The spine may or may not be involved at the same time (spondylodysplasia).

Dwarfism

Dwarfism is defined as a body size below the lowest limit for that age. Differentiate individuals where the limbs are in proportion to the body (proportionate dwarfs), and those where the limbs are not in proportion to the rest of the body (disproportionate dwarfs).

Fever Its presence may indicate an underlying infective or neoplastic process. The nature and course of the fever may give an indication as to the underlying pathology.

Infective processes are typically associated with 'spiking temperatures'.

Postoperative fevers are important signs of complicating factors. An immediate postoperative fever suggests atelectasis of the lung. A fever three days following surgery may indicate phlebitis at a drip site. At five days, a fever is typical of a developing wound infection. At eight days the commonest cause for a fever is a deep vein thrombosis.

Generalised limb or joint disease indicates a widespread problem which may be either inherited or acquired. In many instances this generalised condition is directly related to the presenting complaint of the patient. Bone or epiphyseal dysplasias, or other generalised mesenchymal defects, can lead to stunted growth and limb deformities.

The 'collagen' diseases affect the mesenchymal tissues throughout the body and have widespread effects in many organs. They commonly produce signs in the skin, the joints, the muscles, the bowel and the genito-urinary system.

Colour of the sclerae   Abnormality of collagen structure, such as occurs in osteogenisis imperfecta, allows light to penetrate the sclera, rather than being reflected. This gives the normally white sclera a bluish discolouration.

'Cafe au lait' spots and generalised neurofibromata   suggest the presence of von Recklinghausen's disease, an inherited condition with several musculo-skeletal manifestations. These include neurofibromata and scoliosis of the spine, and congenital pseudarthrosis of the tibia.

Gouty tophi   are typically seen over the big toe, the olecranon and in the pinna of the ear in severe cases of gout. These tophi may be associated with inflammation or deformity of an associated joint, especially that of the metatarso-phalangeal joint of the big toe.

Choreiform or athetoid movements   are involuntary, irregular, incoordinate movements of the limbs, trunk and face. These movements suggests major neurological pathology involving the brain or brainstem and are seen in some cases of cerebral palsy or may follow streptococcal infection associated with rheumatic fever (St Vitus' dance).

Generalised laxity of joints   While often physiological, ligamentous laxity may indicate a condition such as Marfan's syndrome or the Ehrlos-Danlos syndrome.

It may be detected by the ability of the patient to touch his (or her) thumb to the forearm. It is also associated with extension of the metacarpophalangeal joints to 90 degrees, a large carrying angle at the elbows, recurvatum at the knees and the ability to touch the floor with flat hands while standing up.

Ligamentous laxity is often associated with pains and recurrent dislocations of the joints and these can complicate surgical procedures in the affected individual.

Local orthopaedic examination

Examine the local area of the patient's presenting complaint. Specific aspects of the examination is covered in the sections dealing with the spine and each of the limbs. Nevertheless there are some aspects of Orthopaedic pathology which are common to all areas and, for convenience, may be considered at this point.

Look

Look at the skin over the affected area and note the following features:

Texture   Normal skin has a relatively smooth surface which is dulled slightly by the Langer's lines. These lines are indentations in the skin which allow for its normal stretching and contraction. The normal secretion of oils and moisture imparts a slight sheen to the skin. The normal colour of the skin will depend on the ethnic origin of the individual.

Gross local swelling due to subcutaneous oedema will stretch out the skin giving it a tight, shiny surface. When this oedema resolves, the contraction of the stretched skin results in a rough, matt surface for a while before it returns to normal.

Redness Inflammation is a complex reaction to the presence of microorganisms or nonviable irritants. It is readily seen in or under the skin and is an important clinical sign of underlying pathology. It is seen in association with infections, tumours, immune reactions and irritative material which may be endogenous (uric acid crystals) or exogenous (introduced foreign material).

Swelling (oedema) Differentiate localised oedema from generalised oedema. Localised swelling may indicate a localised inflammatory process, as outlined above, or may donate some form of injury – traumatic, chemical, or thermal.

It is common for a limb to swell in the area of injury for a long period after the incident. This applies particularly to injuries of the foot and ankle.

Remember that the outpouring of tissue fluid in an inflammatory situation is rich in protein. This protein may become organised by collagen fibres – leading to fibrosis and scarring of this region with resultant stiffness of the muscles and joints. An important therapeutic measure, particularly where joints are involved, is to maintain movement by means of passive or active physiotherapy once the initial inflammatory process is brought under control. This will help to minimise stiffness and aid the rehabilitative process.

Bruising This is due to the extravasation of blood in the subcutaneous tissues. Direct trauma may produce a magnificent bluish discolouration of the injured skin – the bruise. Dependent tracking of blood, a few days after an injury, is generally of concern to the patient and often indicates fracture of the more bone proximally.

Remember that trauma is relative to the strength of the tissues of the body. Bruising may also occur in conditions of vascular fragility or haemorrhagic diatheses (haemophilia or platelet deficiency).

Lacerations These indicate a significant degree of direct trauma to the skin. A sharp instrument, such as a knife or pane of glass, gives a cleanly incised wound. Ragged wounds result from tearing of the skin, which generally result from blunt trauma or 'high velocity' motor vehicle injuries.

Scars A surgical scar may indicate previous surgery related to the present condition. Ragged scars indicate previous traumatic or infective events which may, or may not, be related to the presenting complaint.

Ulcers   These can indicate previous trauma, an infection of the skin or chronic venous congestion. 'Pressure sores' can develop over the sacrum, heels and trochanters of debilitated, recumbent patients. These ulcers are due to prolonged pressure on a bony prominance and their development is a particularly pernicious complication of therapy. These areas should be regularly checked in paraplegic and elderly, bedbound patients.

Similarly, such ulcers can develop under plaster casts or tight dressings if they are too tight and produce localised areas of pressure on the underlying skin. Heed must be taken of the patient who complains of excessive pain beneath his cast.

Sinuses   Clean, healthy living tissue will not as a rule become infected. A chronic sinus due to a pyogenic infection means that there is a retained foreign body or dead tissue at its base perpetuating the infection. Chronic sinuses may also be due to infection with an organism such as tuberculosis or a fungus.

Callosities and abrasions   These indicate excessive pressure on that portion of the skin and can show up ill fitting footwear or artificial limb prostheses. Callosities under the ball of the foot suggest intrinsic muscle weakness, resulting in excessive pressure on the metatarsal heads and are seen in association with 'hammer', or claw, toes.

Look at the underlying muscle and note the following:

Muscle wasting   Wasting of the muscles is an important clinical finding. Reflex inhibition of muscles, with wasting, commonly occurs after injury and is due to pain. This complication is readily seen in the muscles of the thigh after a knee injury. Other causes of muscle wasting include disease of the muscle itself or damage to its motor nerve.

Fasciculation of a muscle   This occurs after injury to the muscle's motor nerve.

Feel

Use your hands and fingers to detect clinical signs. These examinations should be performed with care as they may cause a greater or lesser degree of pain or discomfort to your patient.

Factors which may be felt in relation to a pathological process anywhere in the body include:

Abnormal temperature   The inflammatory condition not only causes pain, it also causes the local blood vessels in the skin to dilate, increasing the local temperature.

Decreased temperature of the skin usually indicates diminished blood supply to the region. This may be due to cardiac dysfunction or peripheral vascular disease. In the orthopaedic trauma patient it is commonly due to hypovolaemic shock.

Differences in skin temperature are best detected by feeling the area with the back of your hand (as your skin is thinnest here) and comparing it with the opposite side.

Localised dryness or dampness of the skin These symptons indicate changes in the activity of the autonomic nervous system supplying the area. Excessive moisture perhaps indicates nervousness of the patient. Extreme dryness shows loss of autonomic innervation of the skin and may be associated with section of the peripheral nerve.

Tenderness Palpation elicits the sensation of pain, by the patient, in the affected area. The patient can often indicate the painful area, the so called 'point test'. Careful and gentle palpation in this area can reveal the anatomical structure in which the pain is arising. Sprains of a specific ligament of a joint, or fractures of a bone, may be identified in this way.

Crepitus The term crepitus means 'a crackling sound'. This may be heard or felt as an irregular fine vibration in joints, bones or soft tissues.

Crepitus in joints may be due to osteoarthritis where the two irregular bony surfaces, denuded of cartilage, rub together as the joint moves.

Crepitus in bone can be detected when the two rough edges of a broken bone rub together. It is rather unkind to attempt to elicit this sign as it causes severe pain.

Crepitus in tendon sheaths may be felt as the tendon moves within the affected tendon sheath, in patients with 'Tenosynovitis'.

Crepitus in the soft tissues generally indicates the presence of gas. The fine crepitations can sometimes be felt after a compound fracture when air is introduced during the injury or in the chest wall, signifying rupture of the lung from a fracture of a rib. Gas crepitation is also detected in cases of gas gangrene due to Clostridia welchii infection, a serious condition and a major threat to the wellbeing of the patient.

Bony bossing This is common in osteoarthritic joints or may be related to previous trauma. One of the effects of the inflammatory process in osteoarthritis is to cause the affected bones to increase in size. This generally does not cause much of a problem, except in the foot where such enlargement may interfere with the fitting of shoes.

Lumps (abcess, cyst or malignancy) The nature of a localised mass depends on the pathological cause and assessment of the nature of the swelling can give important clues as to the its aetiology.

There are nine points to assess when examining a mass:
- size
- shape
- consistency
- edge

- lobulation
- fixation
- fluctuation
- pulsation
- transillumination.

A localised lump from an inflammatory process is often self evident. However the inflammatory response of a deep seated abcess can be masked by overlying muscles and fascia. A malignant neoplastic swelling is generally fixed by its infiltration, although fixation to the skin or the underlying tissues can also indicate a fibrotic process from chronic infection.

Tumours are generally firm. Fluid filled swellings are soft and fluctuant. There may be blood, serous fluid, synovial fluid, or a clear exudate in cases of simple cysts. A cyst is not generally associated with an inflammatory response. Clear fluid will allow light to pass through; blood or pus will not.

Pulsations, in time with the pulse of the patient, can sometimes be detected with an aneurysm.

Effusion of the joints   Joints normally contain a small quantity of fluid within them for lubrication and nutrition of cartilage. In pathological processes there may be an increase in this fluid or an accumulation of abnormal fluid within the joint. This fluid is detected by swelling and fluctuation of the joint.

Fluid in a joint

Causes of fluid within a joint include degenerative osteoarthritis (a yellowish fluid), trauma (blood stained fluid, which may contain fat globules), infection (a serous fluid or frank pus may be present), rheumatoid arthritis (opalescent yellow or slightly blood stained fluid) and gout (opalescent fluid resembling an infective process). Pus in a joint is extremely irritative and the patient resists any movement of the joint. Extreme reluctance to move a joint, therefore, is very suggestive of an acute septic arthritis.

Synovial thickening   This can only be felt in those joints close to the surface of the skin. Synovial thickening results from conditions within the joint which lead to oedema or hypertrophy of the synovium. Generally these are of an inflammatory nature such as occurs in rheumatoid arthritis, or may be associated with a chronic infection such as tuberculosis, or with blood in the joint such as occurs in haemophilia. More rarely are neoplasms, such as villo-nodular synovitis or synovial malignancies, responsible.

Reflex sympathetic dystrophy   This is a group of related conditions characterised by excessive, prolonged pain in a limb, following some kind of trauma. The area typically is swollen and oedematous, the joints are stiff and the skin is dry and bluish in colour. It is postulated that the condition, in some way, is due to autonomic nervous activity and it often proves difficult to treat.

Move

Move the joints through their normal ranges of movement. Each joint has its normal range of movement, given in the relevant section of this book. A pathological condition affecting a joint may reduce or increase its normal range of movement, depending on whether the constraining structures are contracted or destroyed. Note that any movement in a bone is abnormal.

Position of comfort   Stretching of the capsule, as occurs in an effusion, may produce pain. In this case the joint will be favoured and held in the position that accommodates as much fluid as possible, reducing tension on the capsule – the 'position of comfort'.

Contractures   A 'contracture' is deformity of a joint within its normal plane of movement in which the articular surfaces are relatively normal. It may be fixed or mobile, in which case a greater or lesser degree of movement is possible. Contractures of a joint may arise as a result of one of several possible pathological processes.

Normal joint mobility depends on regular movement of the joint through its full range of movement. Chronic contracture of a hypertonic 'spastic' muscle, or a normal muscle acting against paralysed antagonists, leads to the joint being held in a single position for a long time. Shortening and fibrosis occur in the dominant muscles and tendons. Secondary contractures of the joint capsule then also occur. Organisation of fibrin deposited in and about the joint as a result of inflammation and oedema will add to its contracture.

Conditions affecting the normal movement of a joint include head injuries, and those which cause pain on movement, such as injury and infections of the joint, fractures of a limb bone, and reflex sympathetic dystrophy.

Fibrosis of the soft tissues around a joint will lead to its contracture. Third degree burn contractures of the skin, if over a joint, will cause its contracture. Fibrosis in muscles or tendons after incised wounds can lead to limitation of movement and contracture of a joint. Fibrosis in the muscles, after damage by ischaemia or electricity, give rise to a very contracted, immobile joint.

(Differentiate a muscle contracture from those conditions where there is an intra-articular or extra-articular bony block to full movement, such as a bar or osteophyte.)

Ankylosis   An ankylosis is a fixed deformity of the joint in which the articular surfaces are destroyed. The deformity is usually in the plane of the joint but there may be varus or valgus angulations. It occurs after destruction of the articular cartilage by an inflammatory process and is described as being either 'fibrous' or 'bony'.

Typically the ankylosis results from an infection or a condition such as rheumatoid arthritis, which destroys the articular cartilage. Initially adhesions develop between the bare joint surfaces, and the ankylosis is fibrous, allowing some slight movement of the joint. Later the adhesions within the joint may become ossified, leading to total bony fusion. (When a joint is fused surgically it is termed an 'arthrodesis'.)

Contracted joints

Contracted joints become a great handicap to the patient and make nursing extremely difficult, if this is necessary. In the growing child long standing contracture will cause secondary changes to take place in the joints and bones of the pelvis or limb. The best treatment of contractures is prevention. Prophyllactic physiotherapy in neurological conditions goes a long way to ward off these unwelcome complications. Swollen joints should be regularly moved through their full range of motion, once the initial inflammatory process has subsided. Where contractures have developed, physiotherapy and splinting may help. Often, release of capsule and the offending muscles and tendons is necessary, with or without neurectomy of the relevant motor nerve.

General orthopaedic examination summary

Systematic examination

Examine the following systems in the patient:
   State of health
   State of nutrition
   State of hydration
   Generalised skin colour changes
   Respiratory system
   Cardiovascular system
   Generalised oedema
   Alimentary system
   Genito-urinary system
   Neurological system.

General orthopaedic examination

Assess and record the following general features:
   attitude of the patient
   size of the patient
   fever

generalised limb or joint disease
colour of sclerae
'cafe au lait' spots
gouty tophi
involuntary movements
generalised laxity of joints.

Local orthopaedic examination

Examine the local area of complaint:

Look

Look for or at the following features:
texture
redness
swelling
bruising
lacerations
scars
ulcers
sinuses
callosities
muscle wasting
fasciculation.

Feel

Feel for the following features:
abnormal temperature
moisture of the skin
tenderness
crepitus
bony bossing
lumps
joint effusions
synovial thickening
reflex sympathetic dystrophy.

Move

Assess for the following pathological conditions:
position of comfort
contractures
ankylosis.

3 Examination of the spine

The spine performs two functions. It maintains the trunk in the upright position and it protects the fibres of the central nervous system responsible for carrying motor and sensory impulses between the brain and the limbs. Depending on the level, conditions that affect the spine can affect the function of the arms and legs. Bowel and bladder function are important facets of the body which may also be affected by spinal pathology.

The examination of the spine should be carried out in a warm room which will allow the patient to remove enough clothing for you to see the spine, the abdomen, the arms and the legs. This generally means the patient stripping to the underclothes. Modesty may be maintained by the patient wearing a short gown with the opening at the back.

Examine the shape and movements of the back while the patient is standing. Look from the front, from the side and from the back. Perform the examination of the abdomen, the leg raising and neurological tests while the patient is lying on the back. Palpate the back while the patient is lying on the stomach.

Look

Inspect the skin of the back   Look at its colour and texture, and for the presence of swellings, bruises and lacerations. Pilonidal sinuses are relatively common, other sinuses are rare.

A sinus related to the spine often indicates a congenital maldevelopment of the underlying spine. This may be accompanied by a small area of pigmentation at the base of the spine, with or without a tuft of hair.

Examine the shape of the spine   Examine the spine from the back and from the side. The spine is usually straight when viewed from the back. A tape placed along the spinous processes will show up any deviation from this straight line (Fig 3.1).

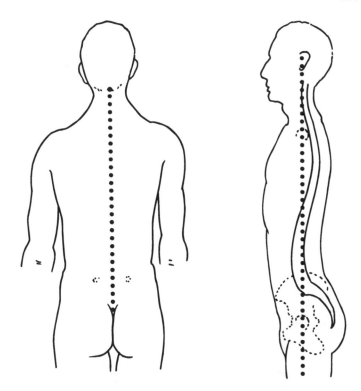

Figure 3.1 Anteroposterior alignment of the spine

Figure 3.2 Lateral alignment of the spine

The curvatures of the spine become evident when the patient stands or sits. When viewed from the side the spine exhibits three curves – a lordotic curve of the cervical spine, a kyphotic curve of the thoracic spine, and a lordotic curve of the lumbar spine (Fig 3.2). However, the overall projection of the spine in this projection is also a straight line.

Stand the patient with his back against a wall   In a normal person the occiput, shoulders, buttocks and heels can touch the wall when the person stands with his back against it.

Sitting ability

Sit the patient on the couch and assess whether he can sit straight, unaided. With a normal spine the body weight is evenly distributed over the pelvis and the posture is erect. Scoliosis or pelvic obliquity can position the centre of gravity of the body over the edge of the pelvis and cause the patient to fall to that side. To prevent this he has to support himself with his hand.

Ability to lie flat

Ask the patient to lie on the couch and put his head back on the pillow. If the neck remains flexed with the head and shoulders off the pillow it is very suggestive of the later stages of ankylosing spondylitis, in which condition the spine becomes fixed in a forward curve.

Deformities of the spine

Note any deviations of the spine from its normal alignment. Small degrees of deformity often go unnoticed but a severe deformity cannot fail to concern the patient, both because of its appearance and as it often limits his functional abilities.

Kyphosis   This is an excessive forward curvature, usually of the thoracic spine, which may be developmental (as in Scheuermann's disease) or due to fracture or dislocation of the vertebrae (Fig 3.3).

Figure 3.3 Kyphosis   **Figure 3.4** Lordosis

Lordosis   This is an excessive backward curvature of the spine, usually occurring in the lumbar region (Fig 3.4).

Scoliosis   This is a lateral curvature of the spine. This common spinal deformity often has a rotatory component to it and can affect the lumbar and/ or thoracic spine. It may be associated with obvious pathology, such as paralysis, but more commonly is idiopathic (Fig 3.5).

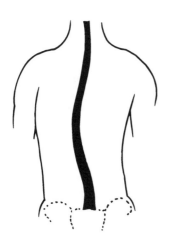 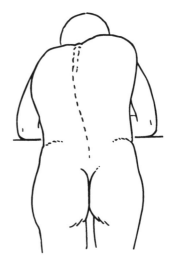

Figure 3.5 Scoliosis   **Figure 3.6** The rib hump

Fixed or mobile deformities

Determine whether the deformities of the spine are fixed or mobile. Observe the patient lying or sitting normally, then passively stretch the spine by pulling the trunk up under the arms. If the curvature is maintained it is 'fixed'; if the spine straightens out it is 'mobile' or 'correctable'. A mobile scoliosis on standing, due to a slight leg length discrepancy will straighten out when the patient sits.

Note the presence of any specific spinal deformities:

Rib hump

Look at the patient from the back   Ask the patient to bend forward. Normally the back remains flat. Long standing scoliosis of the thoracic spine causes the ribs to angulate. These secondary changes may not be evident when the patient stands erect but when the patient bends forward the spine straightens out and the deformed ribs on the convex side become prominent – the 'rib hump' (Fig 3.6).

'Dowager's' hump

Look at the elderly patient from the side   Severe osteoporosis weakens the bone, resulting in wedge compression fractures of the vertebrae. The cumulative effects of these fractures leads to a decrease in height and a kyphotic bowing of the thoracic spine. This is seen particularly in elderly women.

Gibbus

Look at the back for acute angulation   Collapse of two adjacent vertebral bodies causes the spine to angulate forward, producing an acute, visible
angulation at the level of collapse – the 'Gibbus'. A gibbus can result from
fracture of the vertebra(e), infection (commonly tuberculosis) or tumour of
the affected vertebra(e) (often metastatic). The deformity may be accompanied by weakness or paralysis of the legs if the spinal cord is compressed by
the by the lesion.

Feel

Palpate the abdomen with the patient lying on his back   Intra-abdominal
pathology may present as a backacke. Localised swellings from aneurysms
or tumours, underlying the presenting complaint, may be felt.

Radicular signs in the leg are usually due to irritation of the nerve root in
the spinal canal. Occasionally such signs are due to compression of the
sciatic or femoral plexus by a pathological process within the pelvis. A
rectal examination is thus an important part of the examination of the spine.

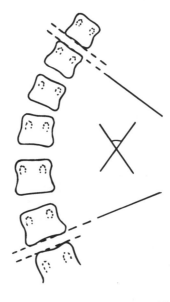

Figure 3.7 Measurement of scoliosis (Cobb's angle). The angle between lines
projected along the end plates of the 'end' vertebrae, as visualised on an
anteroposterior radiograph of the spine. These vertebrae are recognised by parallel
end plates of adjacent vertebrae. The end plates of vertebrae in the curve diverge
from one another.

Palpate the back with the patient on his stomach   A pillow under the hips relaxes the spine and makes it more comfortable for the patient. Palpate each spinous process and the interspinous ligaments firmly. This allows further assessment of the line of the spine and also can indicate whether the pain arises from the spine itself or from structures alongside.

Gaps in the interspinous ligament, following a traumatic dislocation, may be detected by running a finger along the interspinous ligaments.

Palpate both sacro-iliac joints   Feel for tenderness, indicating septic or degenerative arthritis. 'Fibrositis' in the ligaments and muscles in this region is a common source of pain.

Feel the tone of the muscles   Palpate on either side of the spine. Tone is increased if there is pain in the back and the muscles may feel rigid. Tenderness may also be elicited in the muscles.

Feel for areas of increased temperature   These indicate a localised inflammatory process. This is not commonly detected since most inflammatory processes affecting the spine are deep and do not affect the superficial tissues.

Feel for localised masses or swellings   These indicate superficial abcesses or tumours.

Measure

Measure the deformities of the spine   The overlying tissues generally obscure minor spinal deformities. Accurate measurement can only be obtained by using radiographs taken in the lateral and anteroposterior projections. Lines are drawn along the projected surfaces of the vertebrae at the top and bottom of the curve and the angle between them measured. The thoracic spine has a normal kyphosis of less than 20 degrees. There is normally no scoliotic curvature of the spine, and the lines, in this case, are normally parallel (Figs. 3.7–3.9).

Measure for leg length inequalities   The easiest way to perform this is to ask the patient to stand with his feet together and measure the level of the pelvis. If one side is lower than the other there is invariably a mild scoliosis. If so, place blocks under the foot of the shorter leg and determine whether this corrects the spinal curvature. The thickness of the block gives the measurement of the shortening of the leg (see Chapter 5: Examination of the lower limb).

Move

Ranges of movement of the spine

The ranges of movement of the spine are assessed actively as it is difficult for the examiner to move the spine passively.

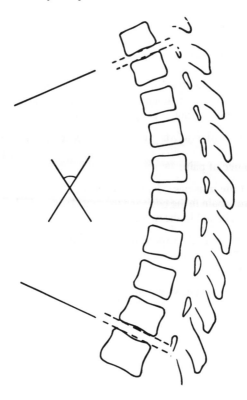

Figure 3.8 Measurement of the angle of kyphosis. The angle between the lines projected from the end plates of the highest and lowest vertebrae, visualised on the lateral chest radiograph.

Neck

The neck normally is very mobile with a wide range of active movement. With flexion it is normally possible to place the chin on the chest. With extension it is possible to see the ceiling. With lateral flexion it is possible to place the ear on the ipsilateral shoulder and with rotation it is possible to place the chin on the shoulder.

Assess the ranges of motion of the neck   Sit the patient on a comfortable chair or examination couch and ask him to forward flex, extend, laterally flex and rotate the neck.

Thoracic spine

The thoracic spine is splinted by the rib cage and clinically there is no measurable movement in this portion of the spine.

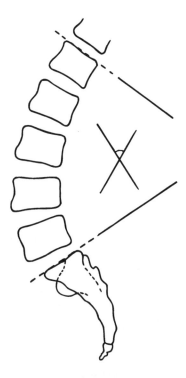

Figure 3.9 Measurement of lordosis. The angle between lines projected along the top of the first lumbar and the first sacral vertebrae.

Lumbar spine

Assess lumbar flexion and extension   Ask the patient to stand and to flex forward, then to bend backward as far as possible. Observe this from the back. There is usually a fair degree of flexion in the lumbar spine, although this movement can be masked by movements at the hip.

Assess lumber lateral bending   Ask the patient to bend to each side as far as possible running his fingers down his thighs.

Assess lumbar rotation   Ask the patient to place his hands behind his head and then turn his trunk as far as possible to either side.

Spinal instability

Ask the patient to bend as far forward as possible and then straighten up. If the spine is unstable the patient feels that he needs support when straightening his back. He places his hands on his thighs and literally climbs up his legs as he straightens up.

Nerve root irritation

Determine whether there is any irritation of the spinal nerve roots. When a nerve root is compressed, by prolapsed disc, osteophyte, ligament or tumour, it becomes inflammed, oedematous and irritable. Stretching of the affected nerve produces pain as it moves across the constriction.

Nerve root irritation can be specifically sought for with various stretch tests:

Straight leg raising test (Lasegue)   Lie the patient on his back and lift the straight leg until the patient complains of pain. Normally the hip can be flexed to 90 degrees. When a lumbar sacral nerve root, supplying the sciatic nerve, is irritated straight leg raising will evoke pain after 25 to 30 degrees. This, particularly if there is a history of sciatica and paraesthesia in the legs, is very suggestive of low lumbar nerve root compression.

Stretch test (Bragard)   Raise the straightened leg until the patient just experiences pain, then lower it slightly. Dorsiflex the foot sharply. This manoeuvre stretches the sciatic nerve and produces pain if one of its roots is inflammed.

Bowstring test   Raise the straightened leg until the patient just experiences pain. Bend the knee slightly at this point to relieve the pain. Push sharply into the popliteal fossa to stretch the nerve lying across it. Again, if the nerve root is inflammed this manoeuvre reproduces the pain.

Femoral stretch test   Lie the patient on his stomach. Pain from an irritated high lumbar nerve root supplying the femoral nerve can be elicited by extending the hip, thereby stretching the femoral nerve.

Sacro-iliac joint inflammation

Assess the sacro-iliac joints in all cases of backache – particularly when there is an element of intra-abdominal pain. Inflammation of the sacro-iliac joints can emulate intra-abdominal pathology and this condition can be mistaken for appendicitis.

Gaenslen test   Lie the patient on his back with the affected side at the edge of the examination couch. Extend the leg over the side of the couch to stress the sacro-iliac joint. If this is inflammed, pain is experienced in the region.

FABER test   (forced abduction and external rotation) Flex and abduct the hip as much as possible on the affected side, externally rotating at the same time. This also stresses the sacro-iliac joint, producing pain if it is inflammed.

Neurological examination

As the spine carries the somatic and autonomic motor and sensory nerve supply to the upper and lower limbs, a full neurological examination of the arms and/or the legs is an integral part of the examination of the spine.

Muscle tone

Assess the tone of the muscles The muscles are hypotonic or atonic (flaccid) in conditions which affect the lower motor neurones or the cerebellum and show increased tone or rigidity with upper motor neurone lesions.

Flacid paralysis develops when there is complete damage to the motor nerve to the muscle (such as when the nerve is injured) or to its anterior horn cell (as occurs in poliomyelitis or motor neurone disease).

Rigidity arising from increased muscle tone occurs in lesions of the upper motor neurones (clasp-knife) or extra-pyramidal system (cog-wheel). The nerves show increased activity in all phases of activity, losing the normal reflex inhibition. Rigidity reduces the power and velocity of muscle action, resulting in marked impairment of function of the limb.

Joint movement

The active movement of the joints is a useful indicator of the motor function of various levels in the cord. The innervation of the upper limb, although in essence a single level for each movement, is more slightly more complex than that of the lower limb. The lower limb has a two-level spinal innervation for each movement.

Upper limb

Shoulder		*Elbow and forearm*	
Flexors	(C5)	Flexors	(C5,C6)
Abductors	(C5)	Extensors	(C7,C8)
Lateral rotation	(C5)	Pronation	(C6)
Adductors	(C6,C7,C8)	Supination	(C6)
Extensors	(C6,C7,C8)		
Medial rotation	(C6,C7,C8)		
Wrist		*Hand*	
Flexor	(C6,C7)	Long flex and ext	(C7, C8)
Extensors	(C6,C7)	Intrinsics	(T1)

Lower limb

Hip		*Knee*	
Hip flexion	(L2, L3)	Knee extension	(L3, L4)
Hip extension	(L4, L5)	Knee flexion	(L4, L5)
Ankle			
Ankle dorsiflexion	(L4, L5)		
Ankle plantarflexion	(S1, S2)		

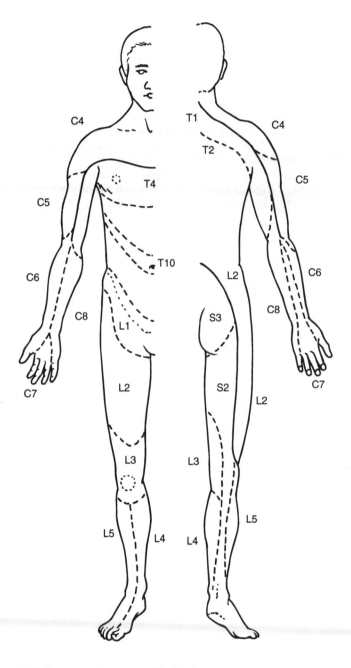

Figure 3.10 Cutaneous dermatome distributions

Motor power

Assess the motor power of the muscles of the limbs. Active movements will indicate the power of the muscles which can be graded from zero to five:

0 no movement
1 flicker of movement
2 movement, but not against gravity
3 movement just overcoming gravity
4 good movement but not quite normal power
5 movement of normal power.

Diminished power indicates muscular dysfunction, generally caused by an interruption in the motor path, either in the central nervous system or in the peripheral motor nerves. However it can also be due to muscle pathology. Diminished motor power should also be differentiated from diminished movement due to local pain. Loss of motor power may be particularly evident in the digits of the hand or foot.

Reflexes

Assess the reflexes for each of the joints individually(elbow, wrist, knee and ankle). A hyper-reflexic limb, especially when clonus is present, indicates an upper motor neurone lesion, whereas a flaccid limb suggests a peripheral nerve lesion.

Muscle fasciculation

Look for fasciculation of muscles This, if present, will indicate a degenerative loss of innervation to the muscle.

Sensation

Sensory loss often indicates the level of injury of the spine, or compression on the spinal cord or nerve roots. While these tests give a gross indication of the level of the neurological lesion, subtle cutaneous loss (and recovery) is best determined by 'two-point' tactile discrimination.

Assess the sensory modalities Test light touch, deep pressure, pain, proprioception, and temperature sensation of hot and cold for each dermatome in the limbs. Ensure that the patient has his eyes closed. The results may be adversely affected if he can see the stimulus being applied.

A rough guide to the dermatome innervation is shown in the Table on page 38. Also, see Fig. 3.10 opposite.

Upper limb

Over clavicle and upper chest	C4
Over shoulder and outer arm to elbows	C5
Radial side of forearm and thumb	C6
Middle fingers	C7
Ulnar finger and side of forearm	C8
Inner arm	T1

Lower limb

Anterior thigh	L1, L2
Anterior knee	L3
Inner lower leg and foot	L4
Outer lower leg and foot	L5
Sole of foot	S1
Back of thigh and lower leg	S2

Autonomic function

Assess autonomic function by the warmth, colour and dryness of the skin. The autonomic nervous system controls the blood supply to the skin and the activity of the sweat glands. Cool, pale, dry skin suggests autonomic dysfunction.

Examination of the spine summary

Look

Inspect the skin for:
    colour
    texture
    swellings
    bruises
    lacerations.
Examine the shape of the spine from the back.
Note any deformities of the spine:
    kyphosis
    lordosis
    scoliosis
    'Dowager's' hump
    gibbus
    rib hump.
Determine whether the deformities are fixed or mobile.
Determine the sitting ability.

Determine the ability to extend the neck.

Feel

Palpate the abdomen.
Perform a rectal examination.
Palpate the back for tenderness and gaps.
Palpate the sacro-iliac joints.
Determine the tone of the muscles.
Feel for localised areas of increased temperature.
Feel for localised masses or swellings.

Measure

Measure the deformities of the spine.
Measure the leg lengths.

Move

Assess the active range of movements of:
  neck
  thoracic spine
  lumbar spine.
Perform the 'stretch' tests
  Sciatic stretch tests – Lasegue, Bragard and 'Bowstring'
  Femoral stretch test.
Assess spinal stability.
Assess sacro-iliac inflammation – Gaenslen and FABER tests.

Neurological examination

Test motor and sensory function of the cord.
Assess the muscle tone.
Motor function:
  measure active joint movement of the limbs
  determine muscle power.
Assess the reflexes – knee, ankle and plantar.
Look for muscle fasciculation.
Sensation – assess dermatome sensory innervation for:
  touch
  pain
  temperature
  vibration
  position sense.
Determine autonomic function.

4 Examination of the upper limb

Examination of the upper limbs requires that you see the patient from the neck to the fingers, on both sides of the body. The shirt or blouse should therefore be removed. Examination should be performed in a warm, private room. Ensure that your hands are warm before touching the patient.

The upper limb in general

Note the shape and position of the arm. The upper limb is generally smooth and well rounded. Its shape will depend largely upon the amount of subcutaneous fat and the muscular development of the individual. It usually hangs, or lies, comfortably down the side of the body with the elbow against the waist. There is a natural outward (valgus) angulation at the elbow – the 'carrying angle'. (This angulation allows items to be carried when the arms are tucked in at the side.) The fingers, gently curved, end at the level of the mid-thigh.

In the relaxed hand the fingers lie together, slightly curved, with the thumb lying beside and turned 90 degrees to the index finger. The position of the fingers depends on the attitude of the wrist. With the wrist slightly extended (the position of function) the fingers are gently curved with the thumb opposing the tip of the index finger. As the wrist flexes the fingers straighten out and the hand opens. As the wrist extends the fingers become more flexed and the hand closes.

Local conditions in the upper limb

Look

Examine the upper limb from the front, from the side and from behind. As outlined in the General section the skin should be examined for: colour changes, inflammation, bruising, lacerations and swellings.

Note any splints or casts on the upper limb. These may be used to aid function of the limb, or to treat a painful condition which may be present.

Note whether any portion of the limb is missing. This may have been as a result of a congenital or acquired lesion. Congenital amputations are rare. Acquired amputations will have occurred as a result of trauma, tumour, vascular injury or infection.

Examine the scapula, shoulder, arm, elbow, forearm, wrist and hand for visible deformities. Is there a deformity of the limb and if so, where is it? Does it lie in a joint or in a bone? (A deformity may be angular, rotatory or in length.)

Is the bone affected?   If so is it angular or bowed? Angular deformities usually indicate malunion of fractures of normal bones. Bowing generally indicates softening of the bones (e.g. rickets, hypophosphataemia, osteogenesis imperfecta).

Is the joint affected?   If so does the deformity lie in the plane of action of the joint or is it away from the line of action of the joint?

Congenital deformities of the upper limb

These include:
Amelia – complete absence of the limb.
Phocomelia – vestigial remnant of the limb.
Ectromelia – partial absence of the limb.
Syndactylism – congenital fusion of the fingers.
Campodactyly – congenital flexion deformity of the fingers.
Gigantism – enlargement of part, or whole, of limb.

Examine particularly the hands, as changes in the hand can proclaim the presence of numerous pathological conditions:

Trophic changes   The hands are in an exposed situation and subject to constant trauma. Callosities on the hands and fingers suggest that the individual engages in heavy manual work. Loss of protective cutaneous sensation leads to trophic ulceration. This is commonly due to neurological sensory deficit such as nerve injury, peripheral neuritis from diabetes, alcoholism, leprosy, or a central demyelinating condition.

Clawing   of some or all the fingers (intrinsic minus hand) suggests ulnar nerve damage (Fig. 4.1). Cicatrisation of the flexor muscles or tendons, following direct or indirect injury, gives rise to fixed flexion contractures of the fingers and wrist.

Wasting of the muscles   Wasting of the hand suggests loss of motor innervation and is most noticeable in the muscles of the thenar emminence and in the interosseous muscles on the dorsum of the hand.

Synovial thickenings in the wrist or fingers joints   These are due to some inflammatory process, commonly rheumatoid arthritis, or occasionally tuberculosis.

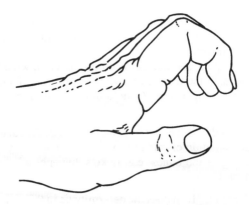

Figure 4.1 Claw hand

Dupuytren's contractures Contractures of the palmer fascia pucker the skin of the palm and cause flexion contractures of the fingers at the metacarpophalangeal and at the interphalangeal joints. These annoying deformities progress until they markedly interfere with the function of the hand.

Hiberden's nodes Osteoarthritis of the distal interphalangeal joints gives rise to small bossings on the phalanges. They may be accompanied by a generalised osteoarthritis.

Tremor Tremor of the fingers may indicate senility, Parkinson's disease or incipient delerium tremens.

Clubbing of the fingers (hypertrophic osteopathy) This may be congenital or can indicate longstanding cardiovascular or pulmonary disease.

Pallor Especially under the nails, pallor may indicate a low level of haemoglobin or hypovolaemia.

Cyanosis This may indicate advanced or acute cardio-pulmonary pathology.

Petechiae under the nails These may also be present over the auxillae and in the sclerae of the eyes in the shock lung syndrome.

Splinter haemorrhages These occur under the nails in septicaemia, infective endocarditis or the shock lung syndrome.

Koilonychia Flattening of the nails indicates a deficiency of iron.

Loss of fingers This occurs in Burger's disease (thrombosing angiitis). This condition leads to progressive loss of the fingers from repeated vascular infarctions.

A single palmer skin crease This is found in patients with Down's syndrome.

Feel

Temperature   Feel the upper limb for any temperature differences. The back of the hand is cooler and has thinner skin, allowing subtle temperature differences to be more easily appreciated.

Moisture content   Feel the skin is for its moisture content. Absence of sympathetic innervation results in a very dry, smooth skin. This is best felt by comparing the two limbs at the same time.

Lumps or swellings   Feel any lumps or swellings in or under the skin. Note their size, shape, consistency, contour, edge, fixation, fluctuation transillumination and pulsations, if present.

Pulses   Feel the pulses in the auxilla, in the cubital fossa and at the wrist. Absence of the pulse may be from an acute event (arterial rupture, intimal damage, thrombosis or embolus) or part of a chronic process (peripheral vascular disease or diabetes).

In cases of diminished or absent pulses assess the viability of the limb clinically, as evidenced by colour, temperature, movement and cutaneous sensation.

Joints   Feel the joints for effusions and synovial thickening. This can be felt in those joints close to the skin, the radiohumeral joint, the wrist joint and the joints of the fingers.

Rheumatoid arthritis typically presents in the early stages with fusiform swellings of the proximal interphalangeal joints as a result of synovitis and resultant effusion.

Fractures and dislocations

Palpate the bony prominences of the upper limb to assess recent fractures and dislocations and their resultant angulations. Displaced fractures may show abnormal movement of the bones and bony crepitus. Dislocations may be detected by abnormalities in the relationships between the bony prominences of the joint.

Both these injuries are associated with swelling and tenderness. They are often obvious from the associated deformity, although minor angular deformities can be masked by the soft tissues. If in doubt, the opposite arm is often normal and available for comparison.

Fractures

Clavicle   The clavicle normally, has a double curve which is palpable subcutaneously along its whole length. Irregularities and displacements of this bone are readily felt.

Humerus   The humerus is subcutaneous only in its most proximal and distal portions where the greater tuberosity and condyles may be felt.

Fractures of its neck or shaft may be obvious from the associated clinical signs and deformity but fractures of the distal humerus are more easily seen and palpated.

The supracondylar fracture of the humerus can be associated with damage to the brachial artery and the distal pulses should always be examined in these cases.

Ulna   The ulna lies subcutaneously and may be palpated along its whole length. Fractures of this bone are usually fairly obvious and clinically detectable.

Fractures of the proximal ulna are commonly associated with dislocations of the radial head – the 'Monteggia' fracture.

Radius   Tenderness over the radial head may suggest fracture. The radius only approaches the skin in its distal portion where fractures are clinically more accessible. The deformity is usually obvious but in some instances there is minimal displacement of the fragments. In these cases the relationship of the radial to ulnar styloid should be assessed. Normally the radial styloid is 1 cm more distal than that of the ulna. The radius often shortens if fractured and displacement of its styloid process implies a fracture.

Wrist   Palpate the wrist joint and the anatomical 'snuffbox' for tenderness of the carpal bones. The clinical examination of this area is important as tenderness in this area may indicate a fracture of one of these bones which may not be obvious on X-ray.

Hand   The metacarpals and fingers are fairly superficial and are usually easily palpated for tenderness and deformity, if the hand is not too swollen.

Dislocations

Sterno-clavicular joint   During traumatic dislocation of the sterno-clavicular joint the proximal clavicle passes either anterior or posterior to the upper sternum. The dislocation is therefore detected by a prominence or hollow at this site.

Acromioclavicular joint   Dislocation of the acromioclavicular joint produces a prominence of the outer edge of the clavicle which can be accentuated by the examiner drawing the affected arm down.

Shoulder joint   The greater tuberosity of the humerus is normally slightly anterolateral to the lateral margin of the acromion. Dislocation of the shoulder medialises the tuberosity, flattens the contour of the shoulder joint and makes the acromion more obvious. The arm typically is held, fixed in abduction.

Elbow joint   The displacement of the olecranon in a traumatic dislocation of the elbow can be medial, lateral or posterior and is usually obvious.. This disrupts the normal, planer, triangular relationship between the olecranon and the medial and lateral condyles of the humerus.

A dislocation of the head of the radius can be seen and palpated lateral to the elbow.

Tenderness and opening of the medial joint line of the elbow with stress may indicate injury or rupture of the medial collateral ligament of the elbow.

Wrist joint   Dislocations about the wrist are often obscured by swelling and may not be obvious, even on X-ray. This particularly applies when only one bone of the wrist is dislocated, usually the lunate. Wrist dislocations also may be associated with fractures of the scaphoid bone.

Hand joints   Carpo-metacarpal dislocations may be palpable but are often clinically obscured by swelling of the hand and may only be detected on careful examination of the X-ray.

It is usually easy to detect a dislocation of the metacarpo-phalangeal or interphalangeal joints of the fingers. Reduction may not be easy if the base of the phalanx has buttonholed through the extensor tendon.

Injury to the ulnar metacarpo-phalangeal ligament of the thumb (game-keeper's thumb) is a relatively common injury and is suggested by laxity of this joint on stressing outward in a 'radial' direction.

Triggering

A localised inflammatory swelling of a flexor tendon prevents it from gliding through the A1 pulley, on the metacarpal neck. This makes flexing the finger difficult. During active flexion, the finger sticks momentarily during its movement, before it suddenly jerks closed. If you hold the metacarpal neck between your thumb and index finger as the patient flexes this finger, a 'click' may be felt as the swelling in the tendon passes through the pulley. Once the finger is flexed, it is often necessary to pull it straight in order to extend it.

Measure

Muscle wasting

Measure the circumference of the upper limbs   Muscle wasting, if gross, is usually obvious. To obtain meaningful measurements the other limb has to be normal. Mark the arms at the same level by taking a point from the same bony prominence on each side. Measure the circumference at this level. A difference in measurement indicates muscle wasting, or underdevelopment, of the narrower limb.

Deformities

Deformities of the limb can be linear (measured in centimetres), or angular, or rotatory (both measured in degrees).

A gentle bowing of a bone suggests soft bone bending under the pull of muscles (e.g. in osteogenesis imperfecta, rickets) rather than the acute angulation of a traumatic or growth lesion. Deformities of a joint may be due to damage or abnormality of its ligamentous, cartilagenous or bony components.

Congenital abnormalities   Abnormalities of the upper limb may take the form of total or partial absence, fusion, duplication or enlargement of the structures. The anomalies may exist in the long axis of the limb, or transversely across it. They may exist in isolation or be part of a generalised congenital syndrome.

In these latter instances, changes in the fingers and hands may be associated with deformities in other parts of the body. Fusion of fingers is found in Apert's syndrome, Arachnoidactyly (long spindly fingers) is typical of Marfan's syndrome.

Acquired deformities   These include shortening, angulation and rotatory deformities resulting from injury, sepsis of the bones or joint, neoplasms or growth plate damage.

Assess the upper limb for deformities in the bones and the joints:

Shoulder girdle   Deformities of the shoulder girdle are best seen from behind. Agenesis and fractures of the clavicle result in the shoulder girdle falling forward and downward. Arrested descent of the scapula results in it remaining high (Sprengel's shoulder).

Shoulder   Because of the wide range of motion of the shoulder, deformities of the upper arm may not be immediately obvious. It is often difficult to detect deformities within the shoulder joint itself.

Elbow   Deformities about the elbow can be seen more easily than those at the shoulder. Angulations at the elbow suggest recent or previous fractures of the distal humerus or dislocations about the elbow joint itself.

A common deformity is a loss of the carrying angle at the elbow (the cubitus varus) arising from malunion of a supracondylar fracture of the elbow in a young child.

Ask the patient to hold the arms horizontally – straight out in front with the palms down. Varus or valgus angulation of the elbow becomes readily obvious.

Forearm and wrist   Angulations of the wrist may indicate pathology of the radius or ulna – trauma, Madelung's deformity or congenital absence. Alternatively, they may be due to damage within the joint itself – trauma, infection, or rheumatoid arthritis (Fig. 4.2).

Hand   Angular deformities in the fingers may be due to injury or degeneration of the joints or tendons. In the later stages of rheumatoid arthritis the metacarpo-phalangeal joints dislocate and are held in ulnar deviation while the fingers show swan-neck or Boutonierre deformities.

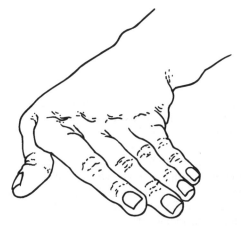

Figure 4.2 The rheumatoid hand. There is subluxation and flexion of the metacarpophalangeal joint, with hyperextension of the interphalangeal joint of the thumb. The metacarpophalangeal joints of the fingers have subluxed volarly. The fingers have deviated towards the ulnar side of the hand. Tendon damage may result in loss of extension of the metacarpophalangeal joints, or a Boutonniere or Swan neck deformities of the interphalangeal joints.

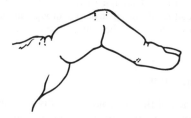

Figure 4.3 Boutonniere deformity – the extensor expansion has split longitudinally and slid down on each side of the proximal interphalangeal joint. When an attempt is made to extend the finger the abnormally positioned tendons case this joint to flex. Overaction of the tendon causes extension of the distal interphalangeal joint.

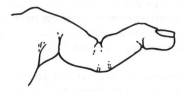

Figure 4.4 Swan neck deformity – subluxation of the proximal interphalangeal joint, or damage to its volar plate, allows it to hyperextend on extension of the finger. Inefficient tendon action allows the distal interphalangeal joint to flex.

A particularly disabling deformity is the rotatory deformity of a finger following fracture of its metacarpal neck. In these cases the affected finger obstructs the others as the hand closes.

Tendon ruptures

Rupture of the rotator cuff will limit active abduction of the shoulder. Ideopathic ruptures of the biceps tendon or expansion can also occur, resulting in a mobile, soft mass in the upper arm when the elbow is flexed. In the fingers, the extensor tendons are prone to rupture, as a result of trauma or from rheumatoid arthritis. Tendon ruptures can cause an angular deformity of a finger joint. A 'dropped finger' can indicate extensor tendon rupture, which is especially obvious with a 'mallet' finger after rupture of the terminal extensor expansion.See Fig 4.5. (Note, however, that a 'dropped wrist' may indicate damage to the radial nerve.)

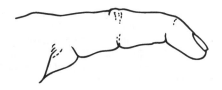

Figure 4.5 Mallet finger. Damage to the extensor tendon over the distal interphalangeal joint allows the distal phalanx to droop. Initially there is no extension of this joint. With partial healing of the tendon the joint shows incomplete extension and an extensor lag.

Length discrepancies

Length discrepancies of the arms are generally less obvious than those in the legs and are more difficult to measure. A small length discrepancy in the upper limb causes no functional impairment.

Contractures

Assess the upper limb for contractures of the joints   Contractures of the upper limb after neurological injury, (particularly after head injuries, strokes and Cerebral palsy) commonly affect the elbow, wrist and hand. These deformities are difficult to treat and impair any limited function which may be present. Contractures may also follow infection or injury to the joint itself.

The shoulder commonly adopts an adducted position, the elbow and wrist a fixed flexion contracture. The deformities of the hand are particularly pernicious. Clawing of the fingers and the 'thumb in palm' make it difficult, not only to use the hand but also to clean it, leading to an offensive smell, distressing to the patient and those around him.

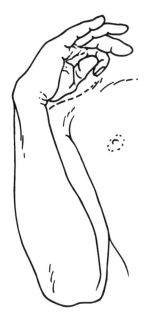

Figure 4.6 Contracture of the upper limb

Move

Range of movements of the upper limb

	Flexion	Extension	Abduction	Adduction	Internal Rotation	External Rotation Opposition
Shoulder	170	45	180	-	90	70
Elbow	150	00	-	-	-	-
Radiohum	-	-	-	-	90	90
Wrist	80	80	-	-	-	-
Thumb	70	-	70	-	-	70
Mc-p	90	40	10*	10*	5*	5*
Ips	90	-	-	-	-	-

* in extension only as the collateral ligaments tighten in flexion

Active movements

Ask the patient to move the joints   (shoulder, elbow, wrist and fingers)
They should be moved as much as possible by himself. The patient should
be able to move the same range of movements actively as the examiner
obtains passively.

Passive movements

Determine the free range of movements of the joints   Do this by firmly, but gently, moving all the joints through as full a range of movement as possible. With experience you can estimate the range of movement but if you are unsure of the angle a goniometer will give accurate measurements

It is preferable for you to perform the passive movements on the normal limb first, if possible. This allows comparison of the two upper limbs and prepares the patient for what is to be done to the affected limb.

Assess the range of movement of the joints

Shoulder

The purpose of the shoulder, in combination with the elbow, is to position the hand for use. Movement at the shoulder is a composite of three movements: thoraco-scapular movement, movement of the gleno-humeral joint and rotation about the clavicle. The functional movements of the shoulder are: flexion, extension, abduction, internal and external rotation.

Flexion (170 degrees)   This is the most common movement, placing the hand in front of the body for use. Loss of flexion and abduction makes it difficult to bring food to the mouth or wash the face, without aids. See Fig. 4.7.

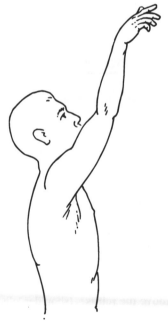

Figure 4.7 Shoulder movements: flexion

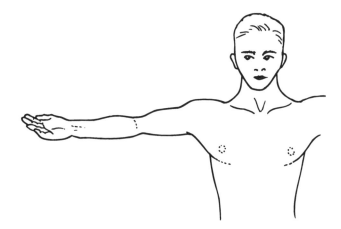

Figure 4.8 Shoulder movements: abduction

Abduction   The arm can be abducted to almost 180 degrees at the shoulder. Note that during this action the scapula allows one degree of abduction for every two degrees of true gleno-humeral joint movement. There is also a degree of external rotation of the humerus during this action (Fig. 4.8). (To assess pure gleno-humeral joint abduction the scapula should be stabilised with your hand.)

Internal rotation (90 degrees)   The elbow is flexed to a right angle and the forearm moved across the chest or up the back (Fig 4.9). Internal rotation is necessary in order to do up buttons and brassieres.

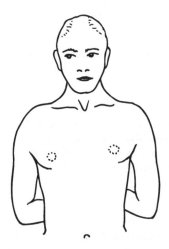

Figure 4.9 Shoulder movements: internal rotation

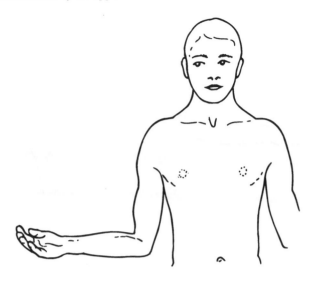

Figure 4.10 Shoulder movements: external rotation

External rotation (70 degrees)   The elbow is flexed to a right angle and the forearm is moved outward from the body (Fig 4.10).

Movements of the shoulder are limited by several common conditions.

'Frozen shoulder'   There is almost complete limitation of movement of the shoulder, actively or passively.

'Painful arc syndrome'   Movement is full but pain is experienced during its mid range, due to lesions of the supraspinatous tendon as they pass under the acromion.

Fractures and osteoarthritis of the shoulder will similarly reduce most movements of the shoulder as they produce pain.

Elbow

The elbow moves only in one plane with flexion and extension.Full extension is zero degrees. The elbow flexes actively and passively to 140 degrees.
 Fractures and dislocations will limit all movements. Bony ankylosis may be the sequel of infection, or rheumatoid arthritis.

Forearm

Supination – (rotates outward) to 90 degrees.
Pronation – (rotates inward) to 90 degrees.

These movements take place at the radio-humeral and distal radio-ulnar

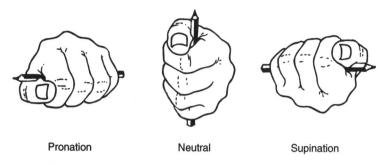

| Pronation | Neutral | Supination |

Figure 4.11 Forearm rotation

joints. Fractures of the forearm and destructive lesions of the radio-humeral and radio-ulnar joints will limit these movements.

Ask the patient to grip a pencil in his palm and measure its excursion from the vertical as the forearm rotates into pronation and supination. See Fig. 4.11.

Wrist

Dorsiflexion – to 80 degrees.
Volar flexion – to about the same.

Angulations from fractures about the wrist and destructive inflammatory arthritis will limit these movements.

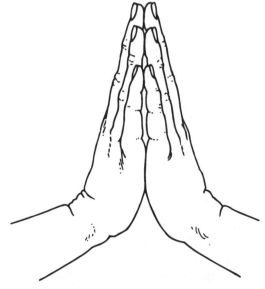

Figure 4.12 Measurement of wrist dorsi-flexion

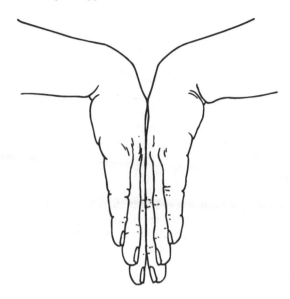

Figure 4.13 Measurement of wrist volar-flexion

Ask the patient to press his palms together and then raise and lower the elbows as far as possible to show dorsiflexion. Repeat this with the back of the hands pressed together to show volarflexion. See Figs. 4.12 and 4.13.

Thumb

Movements of the thumb take place at the carpo-metacarpophalangeal joint and at the metacarpo-phalangeal joints.

Flexion – moves 70 degrees away from the palm.
Extension – moves to neutral.
Abduction – moves to 70 degrees away from the hand.
Adduction – moves 30 degrees into the palm.
Opposition – The thumb rotates into the palm. This movement of opposition is very important and imparts to the hand its tremendous versatility.

Fingers

Metacarpo-phalangeal joints

Flexion – to 90 degrees towards palm.
Extension – to 45 degrees (due to the laxity of the volar plate).

These joints can rotate a few degrees when they lie in the neutral position but not when flexed, as flexion of the metacarpo-phalangeal joints tightens the collateral ligaments, limiting rotation in this position.

Interphalangeal joints

Flexion – to 90 degrees
Extension – 0 degrees (curtailed by the volar plate.)

Full flexion of all these joints places the tips of the fingers in the palm, making a fist. All the finger tips point towards the scaphoid bone when the hand is closed.

Assess tendon function

Assess the function of the tendons to the fingers. All the tests require active movement by the patient.

Flexor tendons

An overall assessment of finger joint and flexor tendon function can be obtained by asking the patient to make a fist. The finger tips normally dig into the palm. Measuring the distance from the tips of the fingers to the palm is one way of assessing a functional deficit. (Fig. 4.14.)

The flexor tendons may then be assessed individually.

There is one flexor tendon to the thumb. Thumb flexion is easy to determine, just ask the patient to bend his thumb.

There are two flexor tendons to each of the fingers, the profundus and sublimis tendons. Assess the function of these separately.

The profundus tendon moves the distal phalanx. Hold the finger extended at the proximal interphalangeal joint and ask the patient to flex the distal interphalangeal joint.

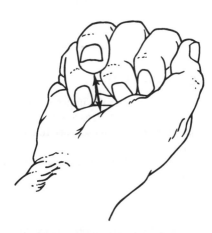

Figure 4.14 Flexion deficit ring finger

The sublimis tendon inserts at the base of the middle phalanx and flexes the proximal interphalangeal joint. To assess sublimis function, the profundus action has to be eliminated. Place the hand supine and hold all the fingers extended, except that to be tested. Ask the patient to flex that finger. The finger will flex at the proximal interphalangeal joint and be completely floppy at the distal interphalangeal joint.

Extensor tendons

Extension of the fingers is a composite action of the long extensors and the intrinsic muscles of the hand. To test pure long extensor function the patient first makes a fist. He then extends the metacarpo-phalangeal joints while keeping the interphalangeal joints flexed.

Interosseous function

One method of assessing the function of the interosseous muscles is to extend the metacarpo-phalangeal joints and alternatively flex and extend the interphalangeal joints. The second method is to keep the fingers straight while flexing and extending the metacarpo-phalangeal joints.

Intrinsic minus (claw) hand

The line of pull of the intrinsic muscles of the hand crosses from the palm to extensor expansion of the fingers. Paralysis of these muscles unbalances the complex relationship between them and the long flexor and long extensor muscles. The interphalangeal joints of the fingers do not extend. To compensate, there is futile overaction of the long extensor muscles. The result is a fixed deformity consisting of extension of the metacarpo-phalangeal joints and flexion of the interphalangeal joints (Fig. 4.1).

Neurological examination

A full neurological examination of the upper limb may be carried out as outlined in the section on the spine but, as a rough check on the innervation of the upper limb, assess the motor and sensory function of the main nerves as a guide to its neurological status

There are two areas of importance when examining the upper limb: the brachial plexus and the hand.

Brachial plexus

The brachial plexus (Fig. 4.15) can be injured by traction during birth, during motor vehicle accidents, or by laceration by a penetrating knife or gunshot

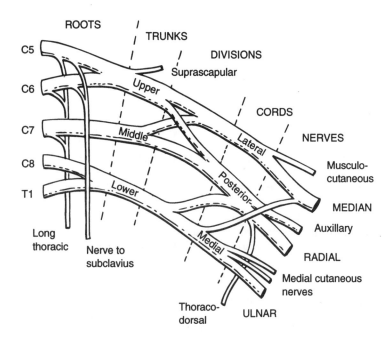

Figure 4.15 Anatomy of the brachial plexus

injury. The classical clinical pictures of upper and lower brachial plexus injury are given by the Erb-Duchenne and Klumpke's paralyses.

Erb-Duchenne paralysis (injury C5,C6)   The arm lies limply by the side, the forearm pronated – the 'waiter's tip' position.

Klumpke's paralysis (injury C8,T1)   This causes loss of intrinsic function and sensation of the hand, resulting in a 'claw hand'.

Injuries to the brachial plexus are often mixed and not as clear cut as this. Only by determining the motor function of the accessible muscles and cutaneous sensation can the site of injury be accurately assessed.

Roots

Nerve to rhomboids   Ask the patient to pull the scapula back towards the midline.

Nerve to serratus anterior   Ask the patient, if possible, to elevate his arm and push against the wall in front of him. Winging of the scapula indicates paresis of this muscle.

Trunks

Suprascapular nerve (from upper trunk only) – Determine contraction in the supraspinatous and infraspinatous muscle with elevation of the shoulder by palpation.

Cords

Lateral cord

Lateral pectoral nerve – Motor to pectoralis major.
Musculocutaneous nerve – Contracts coracobrachialis, biceps and brachialis. Responsible mainly for flexion of the elbow. Provides sensation to the lateral forearm.
Median nerve – The continuation of the lateral cord is joined by a branch from the medial cord to form the median nerve, which supplies motor and sensory function to the forearm and the radial aspect of the palm and fingers.

Medial cord

Medial pectoral nerve – Motor to pectoralis major and minor.
Median cutaneous nerve of arm and forearm – Supplies cutaneous sensation of the anterior forearm to the wrist.
Ulnar nerve – Provides motor impulses to the ulnar flexors of the forearm and most of the intrinsics of the hand. Sensation in ulnar fingers of the hand.

Posterior cord

Nerve to latissimus dorsi – Determine the contraction of the latissimus dorsi when the elbow is pulled into the side.
Auxillary nerve – Motor function to the deltoid muscle.
Radial nerve - Mainly a motor nerve it provides innervation to the triceps, brachioradialis and long extensors of the forearm. Sensory fibres pass to the back of the upper arm and a small area over the dorsum of the wrist and radial aspect of the hand.

Hand

The hand is a precision instrument that plays a vital role in most human activities. It requires a large area of the cerebral cortex to control its movements, and feeds back a large volume of sensory impulses to the brain during its activities. Neurological examination of the hand is vital.

The hand has two main functions: that of pinch movements and that of power grip (grasp). As the hand is a functional instrument these functions should be assessed individually. Ask the patient to pick up, hold and manipulate articles of different sizes and weights.

Pinch movements The precise pinch movements take place along the radial side of the hand, mainly between the thumb and index finger, although opposition of the thumb allows this action with any of the fingers. This opposition should be assessed for each of the fingers.

Power grip This takes place along the ulnar three fingers. (It is these fingers that grip the golf club). The power grip should be assessed by asking the patient to grip the examiner's index and middle finger as hard as possible.

Proprioception and cutaneous sensation The purpose of the arm is to position the hand to touch, feel, hold and move objects. The senses of proprioception and cutaneous sensation are therefore as important as the motor elements in the arms. In addition to the sensations of light touch, pain and temperature, subtle cutaneous loss (and recovery) in the hand is best determined by two point discrimination. Use a pair of dividers or an open paperclip to determine the minimum distance at which the two points can be determined separately, about 2 mm (see Fig. 4.16).

Three nerves are mainly responsible for supplying these functions in the hand:

Median nerve

Motor – Pronates the wrist. Flexes the wrist and the radial fingers. Supplies the short muscles of the thenar emminence to flex, abduct and oppose the thumb, and the radial two lumbricals.

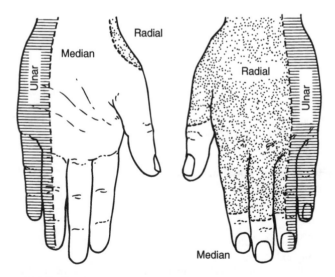

Figure 4.16 Distribution of nerves in the hand

Sensory – Provides sensation to the volar aspect of the forearm, the palm and volar aspect of the thumb, index, middle and half the ring fingers.

Ulnar nerve

Motor – Supplies the ulnar flexors of forearm, the hypothenar emminence, the two ulnar lumbricals, all the interossei (dorsal and medial) and the adductor pollicis.

Test the intrinsic muscles   Ask the patient to hold a piece of paper between the fingers while the examiner pulls it away.

Froment's sign – Loss of adductor function of the thumb causes the interphalangeal joint of the thumb to flex when the thumb and index finger are squeezed together (Fig. 4.17).

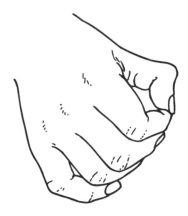

Figure 4.17 Froment's sign. Loss of function of the adductor pollicis, after damage to the ulnar nerve, makes it impossible to keep the thumb straight when squeezing it against the index finger. To do this it is necessary to use the flexor pollicis longus muscle, which flexes the interphalangeal joint of the thumb.

Sensory   Supplies sensation to the dorsal and volar aspects of the little and half the ring fingers.

Radial nerve

Motor – Supplies the long extensors of the wrist, the three long thumb muscles and the long extensors to the fingers.
Sensory – Supplies the back of the arm, the dorsum of the thumb and the first web space.

The ulnar paradox

Cases of isolated ulnar nerve injury result in clawing of the fingers. In cases of high ulnar nerve lesions the two fingers on the ulnar side of the hand, paradoxically, do not claw The reason for this is that long flexor action is necessary to claw the fingers. A high ulnar nerve lesion denervates the muscles on the ulnar side, leaving these fingers unclawed.

Examination of the upper limb summary

Look at the upper limb in general:
  shape
  position.
Examine the upper limb for local pathology.

Look

Look at the skin for:
  colour changes
  inflammation
  bruising
  lacerations
  swellings.
Look for deformities:
  bone
  joint.
Look at the hands for:
  trophic changes
  Dupuytren's contractures
  Hiberden's nodes
  tremor
  clubbing
  pallor
  cyanosis
  petechiae
  splinter haemorrhages
  koilonychia
  loss of fingers.

Feel

Feel for temperature differences.
Feel any lumps or swellings:

size
shape
consistency
edge
lobulation
fixation
fluctuation
pulsation
transillumination
Feel the pulses.
Feel for effusions.
Feel for synovial thickening.
Feel for fractures.
Feel for dislocations.

Measure

Measure for muscle wasting.
Measure deformities:
    linear
    angular
    rotatory.
Measure length discrepancies.
Assess for contractures.

Move

Determine the active and passive range of movements:
    flexion
    extension
    abduct
    adduction
    internal rotation
    external rotation
of all the joints and opposition of the thumb.
Assess tendon function:
    flexor tendons – sublimis and profundus
    extensor tendons
    interosseous muscles.

Neurological examination

There are two areas of importance when examining the upper limb:
    the brachial plexus
    the hand.

Brachial plexus

Erb-Duchenne paralysis
Klumpke's paralysis
Assess lesions of the brachial plexus by the remaining or loss of motor and sensory function in the upper limb.

Hand

The hand has two main functions:
pinch movements
power grip.
Three nerves are mainly responsible for supplying motor and sensation in the hand:
the median nerve
the ulnar nerve
the radial nerve.

5 Examination of the lower limb

Orthopaedic assessment of the lower limb includes assessment of the spine, pelvis, thigh, lower leg and foot. To examine the lower limbs you need to see the patient from the umbilicus to the toes, on both sides of the body. The trousers or skirt therefore should be removed. Examination should be performed in a warm, private room and you must ensure that your hands are warm before touching the patient.

> Abduction – movement of the limb away from the midline.
> Adduction – movement of the limb towards the midline.
> Varus – angulation of the limb towards the midline.
> Valgus – angulation of the limb away from the midline.
> Recurvatum – excessive forward angulation of the knee.
> Equinus – of the ankle. The foot hangs downward.
> Calcaneus – of the ankle. The foot is dorsiflexed.
> Cavus – the foot is excessively arched.
> Pronation – the foot is turned outward.
> Supination – the foot is turned inward.
> See Fig. 5.1

The lower limb in general

The legs are normally of equal length and, in proportion to the body, are about the same length as the torso and head combined. In the relaxed, standing and lying positions the legs are normally turned slightly out, the patella facing about 15 degrees outward, the feet at about 25 degrees.

The feet usually show well developed longitudinal and transverse arches. The medial longitudinal arch is more obvious than the lateral longitudinal arch. The skin under the heels and balls of the feet is thickened but is usually free of blemishes.

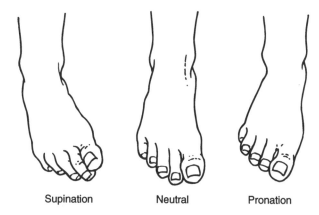

Supination	Neutral	Pronation

Figure 5.1 Foot rotation. The rotary movements of 'pronation' and 'supination' are due to combined movements of the subtalar and midtarsal joints around a longitudinal axis between the centre of the talus and the second toe.

The overall clinical alignment of the legs is along a straight line from the second toe, the middle of the patella, to the anterior superior iliac spine on that side. The knees and ankles normally touch each other on their inner aspect. In order to allow this and to compensate for the width of the pelvis above, the femur normally makes a valgus angle of between 7 and 11 degrees with the tibia (the 'Q-angle' between the thigh and the lower leg).

Radiologically, in both the coronal and sagittal planes, the legs are aligned along an imaginary line passing through the centre of the femoral head, the centre of the knee joint and the centre of the ankle.

Size and appearance   The general size and appearance of the legs and feet should be assessed relative to the other side and to the general size of the body. In some conditions (e.g. neurofibromatosis) the bones, muscles or subcutaneous tissues hypertrophy, symmetrically or asymmetrically.
- Is one leg bigger or smaller than the other in bulk?
- Is one leg longer than the other (relatively or absolutely)?
- Are the limbs in proportion to the body ? (Is the patient a dwarf and, if so, is he or she proportionate or disproportionate ?)
- Are there any localised parts that are proportionately larger or smaller than normal ?
- Are there any parts of the limb that are missing ? Has there been a congenital or acquired (surgical or traumatic) amputation?

Stance

The general stance of the patient gives important clinical information. Normally the body is erect with the weight evenly balanced between each leg.

Muscle weakness, contractures of the joints of the leg, or deformity of the bones or joints will affect the way the person stands and walks.

A pelvic tilt may indicate a spinal problem, a hip problem or a shortening of one of the legs. Excessive lordosis of the lumbar spine indicates a fixed flexion deformity of the hip. Protuberance of the hip, associated lumbar lordosis, may signify an unreduced congenital dislocation of the hip. Deformities of the knees may result in the individual being 'bow legged' or 'knock kneed'.

Walking

The function of the legs is to allow the individual to walk (or to hop, jump, skip, climb, dance..., whatever). The general examination of the lower limbs includes assessment the individual's ability to walk. Walking, or gait, is the movement of the body on the legs as they each pass through two phases: the 'stance' phase, and the 'swing' phase. The stance phase can be divided into three parts; the 'heel strike', the 'flat foot' and 'toe off'.

Walking is really movement due to a series of controlled falls. In thrusting forward on the leg in the stance phase, the centre of gravity of the body is moved forward. The tendency is then for the trunk to fall forward. However the other leg, in the swing phase, now moves in front of the body to catch it before it falls and to take the weight under the centre of gravity as it moves into its stance phase.

Observe the way the patient walks   Gait is normally assessed with regard to the rate of movement, the rhythm of movement, the length of the stride and the cadence of the movement (i.e. the time each leg spends in the stance phase).

Note whether the patient needs an aid for walking   These include canes, crutches, walking rings and calipers. Their use indicates that there is a problem with walking normally as a result of pain, deformity or weakness.

Watch for any limp   Abnormality of the gait is termed a 'limp'. The type of limp can give important information as to the nature of the underlying pathology. Limps may be divided into three main types:
- *The 'antalgic' (or painful) – limp* As little time is spent on the affected leg as possible, as to do so causes pain. This is commonly seen when there is a sprain or arthritis of one of the joints or an injury to the foot.
- *The 'short leg' limp* – Intrinsically there is nothing wrong with the legs except that one is shorter than the other. This causes the individual to dip down slightly on this side whilst walking. The degree of dip will depend on the amount of shortening.
- *The 'neurologic' limps* – A variety of limps due to one of a number of neurological deficits. The neurologic limp is typically due to weakness of the abductor muscles of the hip (Trendelenburg gait); but weakness of the gluteus maximus, weakness of the quadriceps of the thigh or weakness of the foot dorsiflexors (drop foot limp) may apply.

Loss of proprioception in the legs as a result of dorsal column disease or injury will result in a broad based, 'tabetic gait'. The individual compensates for his loss of position sense by widening the stance while walking. Spastic contractures of the hip typically give a scissored, cross-legged gait.

In all these neurological gaits the trunk moves excessively in order to contain the centre of gravity under the legs and prevent falling.

Test for weakness of the hip abductors   The 'abductor' muscles of the hip stabilise the pelvis during the swing through phase of gait. This test generally indicates a neurological disorder of the abductor muscles, but a painful hip can give a similar result due to reflex inhibition of these muscles.

Trendelenburg test   The test is performed by asking the patient to stand. Stand behind him and take the flexed elbows in the palms of your hands. Ask the patient to stand alternatively on each leg.

Normally an individual can stand for a long time on one leg and the pelvis remains level due to the contraction of the hip abductor muscles on that side. If the hip abductor muscles are weak the body will dip to the opposite side and an increase in pressure is felt in your hand.

Local conditions in the lower limb
Look

Observe the lower limb   Observe the limb from the front and from the side and the back - while the patient is standing. Observe it from the front and side while the patient is sitting or lying down.

Note the shape of the leg and foot   Normally they have smooth outlines and contours. The muscles of the leg add to its general appearance. Normally soft when the patient is lying relaxed, the muscles' size and shape will depend on the amount of subcutaneous fat and the athletic interests of the individual.

Note the position of the leg and foot   Intra-articular inflammation may cause an affected joint to be held in a slightly flexed position. The leg will lie in external rotation in fractures of the hip or slipped capital epiphysis. Posterior dislocation of the hip will cause shortening with the thigh held adducted and internally rotated (Fig 5.2).

Inspect the skin for its general colour, texture and integrity   As outlined in the general section, the skin should be examined for: colour changes, inflammation, bruising, lacerations and swellings.

Callosities of the skin on the leg or foot may indicate excessive abrasion from ill-fitting footwear or prosthetic appliances.

Look for signs of muscle wasting   Muscle wasting is most evident in the

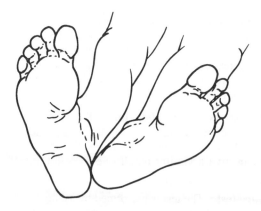

Figure 5.2 Unilateral external rotation deformity of the leg. External rotation of one leg is classically seen after fractures of the hip or leg, or slipping of the epiphysis of the femoral head.

quadriceps muscles as these are the bulkiest of the leg. They are prone to rapid wasting from disuse, commonly after injuries to the knee.

Inspect the joints   The hip is a deep structure and swellings around this joint are difficult to see. The knee and ankle are relatively superficial and swellings around these joints are usually fairly obvious. However it is sometimes difficult to determine whether any swelling is around the joint in the subcutaneous tissues, or within the joint.

An effusion of the knee fills the suprapatellar pouch. When this happens, the slight indentation above the patella is filled out and may become convex instead of concave.

Deformities

Note deformities of the leg   Is there a deformity of the leg and if so, where is it? Does it lie in a joint or in a bone? (A deformity may be angular, rotatory or in length.) Fig. 5.3.

Is the bone affected?   If so is it angular or bowed? Angular deformities usually indicate malunion of fractures of normal bones. The legs take the weight of the body during standing and tend to bow rather than angulate in conditions where the bone softens. (rickets, hypophosphataemia, osteogenesis imperfecta, Paget's disease.)

Is the joint affected?   If so does the deformity lie in the plane of action of the joint or is it away from the line of action of the joint?

Note deformities of the foot   The foot balances and supports the body during walking and is specifically designed to do this. Any deformities of

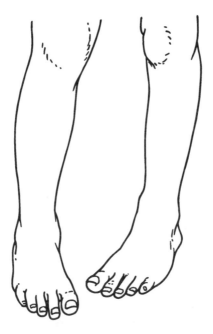

Figure 5.3 Internal torsion of the leg (pigeon toes)

the feet will impair the normal balance of forces within the feet and can seriously impair the individual's ability to walk normally.

Ask the patient to stand and observe the foot from the front and from the side. Flattening or increases of the arches will become apparent. Look at the heels from behind. Normally they are neutral, in the axis of the leg. Varus or valgus angulations will be obvious (Fig. 5.4).

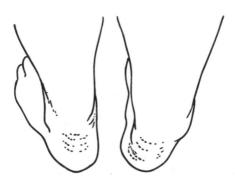

Figure 5.4 Varus heel

Although developmental deformities of the legs in the growing child can be regarded as 'physiological' this is not the case with deformities of the feet. In most cases a much more active approach is required if the child is not to be left with a major incapacity (Fig. 5.5).

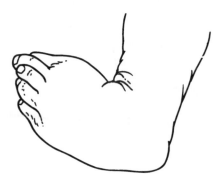

Figure 5.5 Club foot. The foot is plantarflexed (equinus) and rotated medially (varus). Thus the alternative term – 'talipes equinovarus'.

Flat foot   This manifests as flattening of the medial longitudinal arch of the foot. On standing, the foot is turned outward (pronated). The condition is relatively common and flattening of the arches may be 'normal' for that particular person. However, flat feet may also indicate serious bony or neurological anomalies associated with the foot.

Equinus foot   The ankle is fixed in plantarflexion. This may be congenital, often combined with other deformities such as in talipes equinus varus. Children and adults can also acquire an equinus foot from a neurological disorder such as cerebral palsy or cerebrovascular accident.

Cavus foot   This is an excessive longitudinal arch of the foot. The condition is sometimes idiopathic but is more often a manifestation of some neurological condition, such as meningomyelocoele or spina bifida (Fig 5.6).

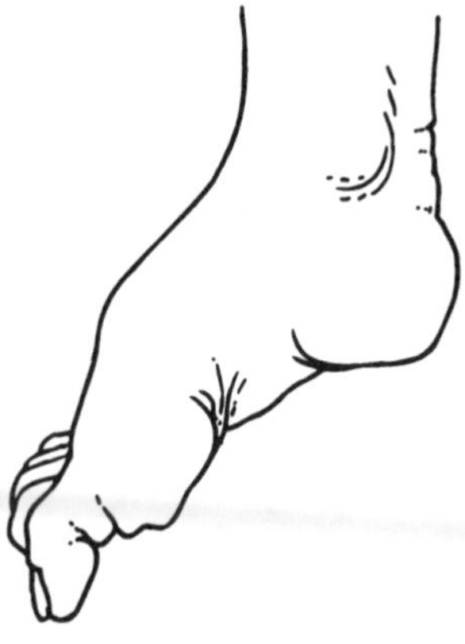

Figure 5.6 Cavus foot. The foot has a high longitudinal arch. It is often held in equinus and has associated clawing of the toes.

Clawed toes   These imply a weakness of the intrinsic muscles of the feet, either acquired, from unphysiological footwear, or as a result of a neurological condition. The toes are responsible for adding support to the ball of the foot during weight bearing. Clawed toes fail to do this, leading to excessive pressure on the metatarsal heads.

Callosities   These develop under the metatarsal heads under the feet as a result of the increased pressure after clawing of the toes (Fig 5.7).

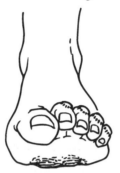

Figure 5.7 Clawing of the toes. Weakness of the intrinsic muscles of the foot allows the four outer toes to claw. This results in loss of support for the metatarsal heads and increased pressure of the skin under them. The skin responds by thickening to give one or more callosities, which are often painful. The flexed toes may develop a callosity on the dorsal surface from rubbing of the shoes.

Bunions   A painful, tender, reddened swelling medial to the first metatarsophalangeal joint signifies increased pressure on the side of the foot, usually from ill-fitting footwear. It is generally associated with a hallux valgus.

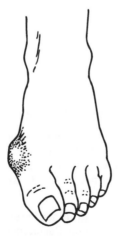

Figure 5.8 Hallux valgus with bunion

Gout   A reddened, inflammed first metatarso-phalangeal joint is typically due to gouty deposits of uric acid in the tissues. These deposits may be severe enough to cause resorption of the bone, and skin ulceration, discharging small white granules. Although gout classically affects the big toe it may affect any synovial joint, commonly the knee or hip joint.

Feel

Palpate the bony contours of the leg   Attempt to determine irregularities or dislocations of the joint or irregularities or fractures of the bones. Tenderness can reveal the anatomical structure in which the pain is arising. (Tenderness of a bone typically donates a fracture.)

The hip   It is difficult to palpate the hip joint but the greater trochanter can usually be felt and its relationship to the iliac crest determined.

The knee   Feel the femoral and tibial condyles about the knee joint. Feel along the joint line itself for its whole circumference for tenderness or irregularity which may indicate capsular, collateral ligament or meniscal injury.

The patella   Feel the patella and note its position. It lies between the femoral condyles in the centre of the joint. Displacement from this axis indicates a dislocation of the patella which may be acute, recurrent or habitual.

The 'apprehension' test   Cases of recurrent dislocation of the patella are revealed by the expression of apprehension on the face of the patient as you try to push the patella laterally.
   The lower pole of the patella lies just above the anterior joint line. It sometime lies more proximal than this – the patella alta. Palpate beneath the surfaces of the patella. This does not usually cause pain except in cases of fracture or in chondromalacia patella.

The tibia   Palpate along the anterior border of the tibia which lies subcutaneously. Although the tibia itself is straight this border normally has a very slight bow to it.

The ankle   Feel the lateral and medial malleoli of the ankle. Being subcutaneous they are easy to feel. The lateral malleolus lies behind the medial malleolus along an axis of about 25 degrees to the coronal plane of the tibia. Bony tenderness may confirm the diagnosis of a fracture.

The foot   Palpate the dorsal surfaces of the tarsal and metatarsal bones of the foot. They are superficial and usually lie on a flat plane longitudinally and slightly curved laterally. As in the hand and wrist, dislocations of the small bones of the foot may not be readily palpable, especially if there is any degree of swelling present.

Feel any lumps or swellings in or under the skin   Assess their size, shape, consistency, contour, edge, fixation, fluctuation and transillumination. Note any pulsations.

Feel the pulses   These should be present in the groin, in the popliteal fossa and in the foot, over its dorsum and behind the medial malleolus. Absence of a pulse may be from an acute event (arterial rupture, intimal damage, thrombosis or embolus) or part of a chronic process (peripheral vascular disease or diabetes).

In cases of diminished or absent pulses assess the viability of the leg clinically, as evidenced by colour, temperature, movement and cutaneous sensation.

Feel the affected portion(s) of the leg   These will be indicated by the patient as being a problem and any of the following may be detected:

Abnormal temperature   Differences in temperature can be assessed by comparing both sides. There is normally a gentle temperature gradient from the cooler foot to the warmer hip. The hand is passed slowly up the leg from the foot. Any sudden, localised temperature increases are very obvious.

Joint line tenderness   The joint lines should be gently palpated along their length. Localised tenderness is present with sprains of the collateral ligaments of the knee or of the ligaments of the ankle. Joint line tenderness may also indicate fracture of a condyle or meniscus, osteoarthritis, ligament injury or, peculiar to the knee, meniscal damage.

Swellings   In the subcutaneous tissues and bone swellings should be assessed for their size, shape, consistency, edge, lobulation, fixation, fluctuation, pulsation, and transillumination.

Gaps   Gaps of a ruptured quadriceps or patellar tendon, or a rupture of a muscle are fairly obvious.

The gap test   Run a finger over the anterior part of the patellar and quadriceps tendons, below and above the knee, pushing down firmly. Normally the finger moves in a smooth, continuous line from the tibial tubercle to the thigh muscles. If the finger dips down into a depression there is a rupture of this part of the extensor mechanism of the knee.

Effusions of the joints   The knee and ankle joints are the easiest to assess for the presence of intra-articular fluid as they are relatively superficial. It must be remembered that similar extravasations of fluid may occur in any of the joints of the leg.

Patellar tap test   This is the most appropriate test to determine the presence of fluid (synovial fluid, blood or pus) in the knee. Fluid, if present, is expressed down from the suprapatellar pouch into the lower knee joint using the first web space of one hand. This fluid distends the lower portion of the knee joint and causes the patella to rise. The index and middle finger of the opposite hand push sharply down on the patella. The patella will be felt to move down and knock sharply against the femoral condyles (see Fig. 5.9).

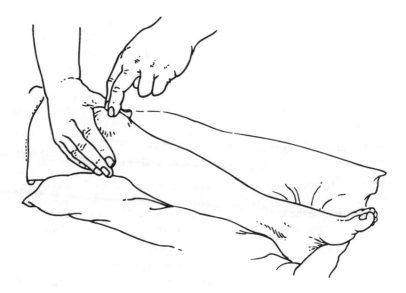

Figure 5.9 The patellar tap test

Bony bossing   Osteoarthritis of a joint often stimulates the development of new bone, both by enlarging the condyles and by producing osteophytes. This overgrowth of bone near the joint can be felt as bony 'bosses' or protuberences, best felt around the knee.

Synovial thickening   This can be felt in the knee, ankle or joints of the foot. This is associated with chronic inflammatory processes taking place within the joint such as infection, rheumatoid arthritis or haemophilia. These processes stimulate the hypertrophy and subsequent thickening of the joint synovium. Villo-nodular synovitis, a tumourous condition of the synovium similarly causes synovial hypertrophy. Malignant tumours of the synovium are rare but will also produce thickening of the synovium.

Crepitus   This may be detected in the bones, indicating a fracture, or in the joints, indicating severe osteoarthritis. This is may commonly be felt in osteoarthritis of the hip or knee. Occasionally the grating of bone on bone may be heard when the patient walks or moves his joint.

Measure

Muscle wasting

Measure the circumference of the lower limbs   Muscle wasting, if gross, is usually obvious. In order to obtain a meaningful measurement the other limb has to be normal. Mark the legs at the same level by taking a point from the same bony prominence of the knee or ankle on each side. Measure the circumference at this level. A difference in measurement indicates muscle wasting, or underdevelopment, of the narrower limb.

Angular deformities – varus and valgus

In the anterior/posterior projection, the legs lie alongside each other, with the condyles of the knee and medial malleoli of the ankles within a centimetre or two of each other.

Assess any deviation from the central line   and then measure the angle in degrees or the displacement in centimetres. You must assess whether the deformity is in the long bone or in the joint.

Joint deformity

The deformity of a joint may be in any plane (linear, angular and rotatory) and may be either fixed or mobile. Joint deformity may indicate previous injury, infection, idiopathic or inflammatory arthrosis, epiphyseal dysplasia or a neurological problem.

During childhood the legs are not static in their appearance but undergo a series of twistings and angulations, in both the hip and knee, as the child grows. These 'deformities' can be a source of acute anxiety to the parents but are regarded with equanimity by the infant, who generally outgrows them.

In infancy the knees have a tendency to bow (genu varum). As the infant develops into a toddler and starts to stand and walk the legs become knock-kneed (genu valgum deformity).See Fig. 5.10.

Measure the distance   between the two medial malleoli at the ankle in valgus deformities, or between the medial femoral condyles in varus deformities. This gives a measure of the deformity and allows one to plot the course of the deformities over several months or years. In most cases these angulations have corrected themselves at the end of growth and can be safely observed.

Rarely do they reach the distance of 8 centimetres nor persist past the age of 8 years. If they do so they should be considered pathological and action taken. Genu varum (bowed legs) deformities in later childhood and adolescence should be regarded as pathological. These will include cases of Blount's disease.

Genu varum or genu valgum deformities in the adult occur in knees which previously have been normal. In these cases the deformities may follow a fracture, which has depressed one of the condyles of the knee or more commonly it develops with osteoarthritis of the knee, commonly on the medial side.

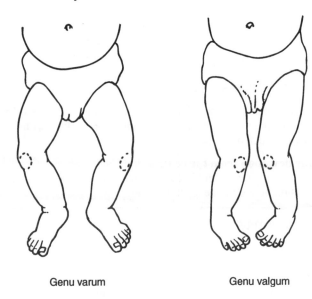

Genu varum   Genu valgum

Figure 5.10

Bone deformity

The deformity of a bone may be in any plane (linear, angular or rota-tory) and may be either fixed or mobile. The long bones are normally relatively straight so that any angulation, or bowing, is pathological. Deformaties may indicate previous injury, infection, growth plate damage congenital or metabolic bone dysplasia or softening of bone.

Rotatory deformities – tibial and femoral torsion

An in-toed gait (pigeon toed) hinders the individual from walking and run-ning as he tends to trip over his feet. It can be due to excessive anteversion of the femoral neck, or to internal tibial torsion.

Measure the internal rotation of the hip   The measurement is done in one of three ways:
1. The patient lies supine and the straight leg is simply rolled inward. The arc of movement of the foot is noted.
2. The hips and knees are flexed to 90 degrees and then turned away from the midline. The arc of movement of the lower leg, from the neutral position, gives the degree of internal rotation of the hip.
3. The patient lies prone and the knees are flexed. The lower legs are moved from the vertical away from the midline and the arc of movement gives the measurement of internal rotation.

If the movement of internal rotation is excessive (i.e. > 20° it suggests that there is femoral anteversion.

Measure internal torsion of the tibia   Seat the patient with the legs hanging over the edge of the bed. Internal version of the tibia can be seen and measured by the degree of internal rotation of the malleoli. (The normal inclination of the transmalleolar plane is at about 25 degrees to the coronal plane.)

Leg length discrepancies

The legs normally are of equal length and equal size. Discrepancies may arise from congenital causes such as poor development of part, or the whole, of the leg; or it may be acquired from conditions affecting the growth of the leg; or from fractures or bowing of the leg.

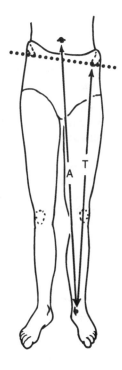

Figure 5.11 Leg length deficiencies.

A – the 'apparent' leg length. Measured from the umbilicus to the medial malleolus. This measurement takes no account of pelvic tilt nor contractures of the hip.

T – the 'true' leg length. Measured from the anterior iliac spine to the medial malleolus. This measurement gives a more accurate assessment of the leg length.

A factor that significantly affects the apparent length of the legs is a contracture about the hip. This deformity should be sought for whenever there appears to be leg length discrepancy. An adduction contracture will seemingly shorten the leg while an abduction contracture will seemingly lengthen the leg.

Measure the difference between the levels of the medial malleoli   This should be done using a tape measure with the feet together. This method gives the measurement of the leg lengths relative to each other. This measurement does not accurately compensate for pelvic tilt or hip contractures.

Measure from the umbilicus to the malleoli on either side   This gives the apparent lengths of the legs. This method suffers the same shortcomings as the relative measurement. See Fig. 5.11.

Measure the legs from the anterior iliac spines to the medial malleolus   Again use a tape measure. This method overcomes the effect of hip position and will give a fairly accurate measurement of the true leg lengths.

Stand the patient and assess the level of the pelvis from behind   Use the hands or a level across the iliac crests. Shortening of a leg is given by the thickness of the block necessary to bring the pelvis back to the horizontal.

Determine whether the shortening is in the femur or the tibia. Measure and compare the distance between the tip of the greater trochanter and the iliac crest. A difference in length between the two sides suggests that the shortening lies in the femoral neck (coxa vara or fracture) or hip joint (osteoarthritis or avascular necrosis).

Bend the knees to a right angle and place the feet together on the examination couch. Look at the level of the femoral condyles from the side. If the condyles of one femur are in front of the other, that femur is longer. If one knee is lower than the other, that tibia is shorter than the other.

Figure 5.12 Hip movement: flexion

For confirmation and absolute accuracy of leg length measurements, a radiographic scannogram using conventional X-rays or a CAT scanner should be performed.

Move

Move the legs actively and passively through the normal range of movements of all the joints.To assess disorders of joint movement, their normal range of movement must be known. The range of movement is given in degrees. (Movement of the bone is always abnormal.) (See Figs. 5.12–5.15)

If you have normal joints and are in doubt as to the range of motion of the patient's joint, surreptitiously compare it with your own.

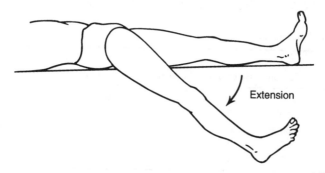

Figure 5.13 Hip movement: extension

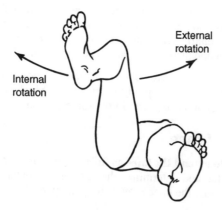

Figure 5.14

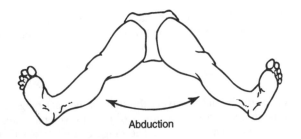

Abduction

Figure 5.15

Movements of the joints of the lower limb

Ranges of movement (degrees)

	Flexion	Extension	Abduction	Adduction	External rotation	Internal rotation
Hip	140	10	45	20	40	20
Knee	140	0	0	0	5	5
Ankle	60	20	0	0	0	0
Subtalar	0	0	5	10	0	0
Midtarsal	5	5	5	5	10	10

Passive movements of a joint   You have to move the joint for the patient and put it through as full a range as is possible. Stabilise the pelvis or the leg above the joint with one hand before assessing its range of movement with the other.

Care must be taken to cause as little pain as possible with this manoeuvre. An unexpected movement of a painful joint can disturb the patient. It is useful to move the unaffected joint first as this prepares the patient for what is about to happen and also allows you to compare both limbs.

Active movements of a joint   These indicate the power of the relevant muscles and will also show up any extensor lag which may be present. Refer to the chart on muscle power in the spine section.

Extensor lag of knee

Ask the patient to extend the knee as fully as possible   If the knee is not straight take the leg and determine whether you can extend the knee any further. The angle that the knee moves, if at all, gives the 'extensor lag'.

Contractures

Determine any contractures of the joints.

The Hip

Flexion contracture   For the Thomas test lie the patient supine. Place one hand behind the lumbar spine of the patient. Flex the 'normal' hip with your other hand until the lumbar spine is felt to have flattened. The degree to which the other leg rises off the couch indicates the degree of fixed flexion deformity (Fig. 5.16).

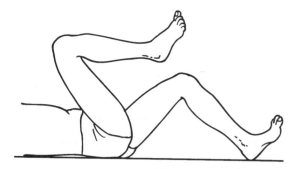

Figure 5.16 Thomas's test. The affected hip flexes as the other hip is passively flexed in order to flatten out the lumbar lordosis.

A spastic flexion contracture of the iliopsoas will cause pain in the hip. The hip not only flexes but the femoral head tends to sublux and, if untreated, the femoral head may finally dislocate; particularly in the child. A flexion deformity of the hip will also cause an increased lordosis of the lumbar spine when standing.

Adduction contracture   Lie the patient supine and span across the anterior superior iliac spines with one hand. Grasp the leg to be tested by the ankle and note its position relative to the pelvis. The hip is then abducted. In the event of an adduction contracture the pelvis moves out with the leg.

Unbalanced action of the hip adductors, due to a neurological disorder, often produces an adduction contracture. Another common cause is prolonged pain in the hip. Osteoarthritis or other inflammatory condition in the hip cause the individual to hold it in a protected position – flexed and adducted. Lack of movement will allow a fixed contracture of the hip to develop. Effectively this causes an apparent shortening of the leg on the affected side. This debilitating deformity often requires division of the adductor tendons to restore adequate movement of the hip.

Abduction contracture   Lie the patient supine and span the anterior superior iliac spines with one hand. The other hand grasps the leg and the position of the leg is noted. The hip is then adducted. In the case of an abduction deformity the pelvis moves with the leg.

Contracture of the hip abductor muscles or of the fascia lata, extends over both the hip and knee. Straightening of the flexed knee will cause abduction of the hip. In these cases there will not only be a flexion, abduction deformity of the hip but a flexion, valgus deformity of the knee.

Contracture of the tensor fascia   Lie the patient prone and extend the knee of the leg to be tested. As the fascia lata passes over both joints and, as the knee is extended, the hip flexes and the pelvis rises off the examination couch.

The knee

Hamstring contracture   Seat the patient with the legs over the edge of the examination couch. Extend the knees. As the knees straighten they come together. Ask the patient to stand and touch the floor. It is not possible for him to bend forward very far. Hamstring contracture causes the pelvis to tilt backward and the knees tend to flex during walking. (Note: spondylosis of the lumbar spine in an adolescent can cause spasm of the hamstrings.)

Quadriceps contracture   Lie the patient supine with the legs extended. Flex the knee of the leg. There will either be a block to full flexion or, more commonly, the patella will dislocate laterally (recurrent or habitual dislocation of the patella).

The ankle

Tendo achilles contracture   Lie the patient supine and dorsiflex the foot with your hand. It is not possible to fully dorsiflex the ankle and the calcaneus may move into varus as you move the ankle.

As the gastrocnemius extends over both the ankle and the knee gastrocnemius contracture may be differentiated from soleus contracture. Hold the foot in as much dorsiflexion as possible and flex and extend the knee. If the contracture is isolated to the gastrocnemius the ankle will plantar flex as the knee is extended.

Tendon ruptures

Tendon ruptures in the leg mainly affect the quadriceps tendon and the tendo achilles.

Rupture of quadriceps femoris expansion   Ask the patient to extend the knee while seated on a chair or couch. Rupture of the quadriceps expansion (above, below or through the patella) is revealed by the inability to extend the knee actively. Passive movement is normal.

Rupture of the tendo achilles Lie the patient prone and ask him to actively plantarflex the ankle against the resistance of your hand. Often this is not possible because of pain.

Lie the patient prone. Grasp the mass of the gastrocnemius muscle and squeeze it with your hand. If the tendon is intact the foot plantarflexes.

Stability of the joints

Although the joints of the leg allow movement of the limb they are very specific in the ranges of movement that they allow. The hip exhibits a very wide range of movement in all planes, while the knee and ankle allow movement in one (saggital) plane only. Within these ranges of movement the joints are relatively stable, the stability being conveyed by the configuration of the bony surfaces and by the restraining ligaments.

Note any excessive movement of the joints This may indicate a disorder of collagen such as osteogenesis imperfecta, Marfan's syndrome or the Ehrlos-Danlos syndrome. In these conditions the collagen structure is deficient. A congenital dislocation of the hip in these conditions is uncommon but pain, a feeling of instability and increased laxity may occur in the knee or ankle at any age. Surgery is not helpful in these cases.

Abnormal movements of a joint indicate a loss of stability of that joint. This may have been due to damage of the bone due to injury, ischaemia or tumour; to damage of the restraining ligaments by injury, Rheumatoid or infective erosions; or to loss of the articular cartilage from excessive loading or an inflammatory or infective process.

Hip

Telescoping

Lie the patient supine Stabilise the pelvis with one hand; the other hand grasps the flexed knee. Pull and push on the knee. Movement of the hip in a longitudinal direction indicates a subluxation or dislocation of the hip with severe instability.

Ortolani's and Barlow's tests are specific for congenital dislocation of the hip in neonates, where the femoral head is either reduced into the acetabulum or dislocated from the acetabulum respectively. These tests become negative after about 3 months, even if the hip remains dislocated.

Ortolani's test Lie the infant supine. Stabilise the pelvis with one hand and grasp the knee between the thumb and index finger of the other. The knee and hip are flexed and the hip is abducted. If the hip is dislocated it is felt to relocate with a palpable click as it passes over the posterior rim of the acetabulum. See Fig. 5.17.

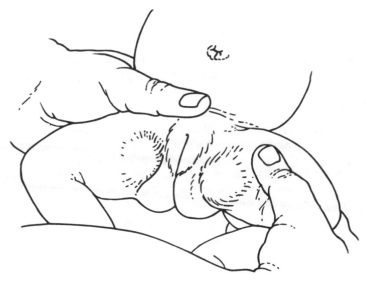

Figure 5.17 Ortolani's test

Barlow's test   Lie the infant supine. Stabilise the pelvis with one hand and grasp the knee between the thumb and index finger of the other. The hip and knee are flexed and the leg adducted while pushing longitudinally on the femur. A palpable click is elicited as the femoral head dislocates over the posterior acetabular rim. See Fig. 5.18.

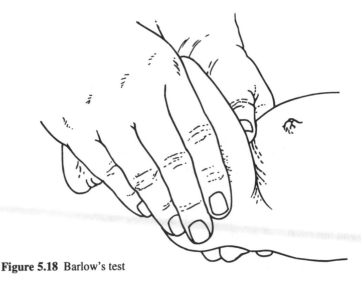

Figure 5.18 Barlow's test

Knee

The knee is very dependent on its restraining ligaments for its stability. These are the medial and lateral collateral ligaments and the anterior and posterior cruciate ligaments. Each ligament should be tested individually.

The most common cause for instability of these ligaments is a traumatic sprain or rupture, usually during some sporting activity. Chronic ligamentous laxity of the knee gives rise to sensations of giving way of the knee during activities. Laxity may not be apparent immediately following injury, due to spasm of the muscles around the knee. An examination under anaesthesia may be necessary in this situation. In the 'chronic' situation the supporting muscles have relaxed and the instability is immediately apparent.

Test for collateral ligament damage

Stress test   Test the integrity of the medial and lateral collateral ligaments of the knee. Flex the knee slightly and stress the joint to the opposite side, assessing the degree of opening of the joint.

Minimal opening indicates a normal ligament. Wide opening suggests a sprain or tear of the ligament. The 'end point' to the stressing is important in assessing sprains. A solid 'end point' indicates an intact ligament in good condition (a grade 1 sprain). A soft 'end point' suggests a ligament which has been stretched and is soft (a grade 2 sprain). A joint which has no 'end point' suggests that the ligament has been completely ruptured (a grade 3 sprain).

Test for cruciate ligament damage

Recurvatum test   The anterior cruciate ligament prevents the knee from bending too far forward. Suspend both heels in the air. Excessive anterior angulation of one knee suggests a torn anterior cruciate ligament on that side. (Females often have a greater degree of recurvatum than males. This angulation affects both knees equally.)

Posterior sag test   Flex both knees and place the heels together on the examination couch. A falling back of one knee suggests rupture of the posterior cruciate ligament on that side. See Fig. 5.19.

Anterior and posterior drawer tests   Flex the knees to 90 degrees and place the heels together on the examination couch. Put your buttock on the toes of the leg to be examined. Grasp the knee with both hands, placing your fingers on the tendons of the hamstring muscles to ensure that they are lax. Move the knee back and forth. Any laxity of the joint is immediately apparent. An anterior shift of the tibia indicates anterior cruciate rupture whilst a posterior shift of the tibia shows posterior cruciate rupture. See Fig. 5.20.

Lachman test   In acute injury the knee is very painful and the patient will not allow sufficient flexion to perform the 'drawer' tests. Lie the patient

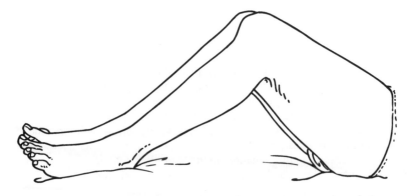

Figure 5.19 Posterior sat test

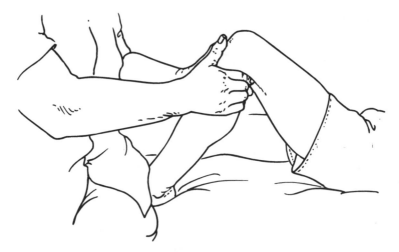

Figure 5.20 The anterior drawer test

supine and flex the knee to about 30 degrees. Grasp the thigh above the knee with one hand and the upper tibia with the other. Move the knee back and forth. Displacement of the tibia indicates which cruciate ligament is lax.

'Pivot shift' test This indicates a rotatory subluxation of the knee with damage to both the antero-lateral capsule and the anterior cruciate ligament. Lie the patient on the unaffected side and apply a valgus strain to the knee as it is alternatively flexed and extended. As the knee passes a critical point at 10 to 20 degrees of flexion there is a palpable 'click' as the torn anterior cruciate allows the tibia to sublux on the femur.

Test for meniscal damage

Varus or valgus strain on the knee, particularly when accompanied by rotation can trap the meniscus between the femoral and tibial condyles and tear it.

McMurray test   This is for the medial meniscus. The patient lies supine. Flex the knee to 90 degrees and hold the thigh with one hand. Hold the heel with the other. Extend the knee with a varus strain and slight rotatory movements. A torn meniscus will become caught between the condyles, causing pain in the knee. A flexed knee indicates a tear of the posterior horn, an extended knee indicates an anterior horn tear. For the lateral meniscus the test is repeated, this time putting a valgus strain on the knee. See Fig. 5.21.

Grinding test   Again this is a test for tears of the meniscus. The patient lies prone. Flex the knee to a right angle. In this position the ligaments of the knee are lax but the meniscus is compressed. The knee is then extended, using a rotatory movement to grind the femoral and tibial condyles together, catching the meniscus between them if it is torn. If the meniscal tear is in its posterior portion the patient will experience pain when the knee is flexed. If the tear is anterior, the pain will be felt as the knee is extended.

Heel-shake test   Hold the heels and extend the legs just above the examination couch. Shake the legs gently at first and give a single, violent shake. Pain in the knee suggests meniscal damage.

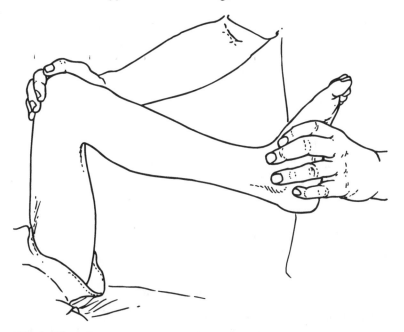

Figure 5.21 McMurray's test

Neurological examination

The motor and sensory function of the major nerves supplying the lower limb should be assessed as outlined in the section on the spine. Motor power, proprioception and cutaneous sensation are important in all areas of the lower limb. Of relevance to the leg, the lumbar and sacral plexes should be assessed.

The lumbar plexus

The lumbar plexus lies within the psoas muscle and is formed from the anterior primary rami of the four lumbar nerves. The main branches which supply the lower limb are from its posterior divisions:

Lateral cutaneous nerve of thigh

Sensory – To the lateral side of the thigh to the knee.

Femoral nerve

Motor – Flexes the hip and extends the knee. Supplies iliacus, pectineus, sartorius, the three vasti and rectus femoris.

Sensory – Upper medial and anterior thigh to the knee. The saphenous nerve continues on to supply the medial aspect of the lower leg and foot.

Obturator nerve

Motor – Adducts the hip. Supplies the obturator externus muscle of the hip and the adductors brevis and longus.

Sensory – Lower medial thigh

The sacral plexus

The sacral plexus lies in front of the piriformis muscle. Its nerves divide into anterior and posterior divisions.

Anterior divisions

Nerve to obturator internus

Motor – Obturator internus and superior gamellus.

Nerve to quadratus femoris

Motor – Inferior gamellus and quadratus femorus.

Medial popliteal nerve (via sciatic)

Motor – motor to the popliteal fossa. It supplies the plantaris, both heads of gastrocnemius, soleus and popliteus. It continues to supply the small muscles of the sole of the foot (medial and lateral plantar nerves).

Sensory – posterior leg (via sural nerve) and the skin of the sole of the foot (medial and lateral plantar nerves).

Posterior divisions

Sciatic nerve

Motor – Flexion of the knee and dorsiflexion of the foot. Supplies the hamstring muscles and the ischial fibres of the adductor magnus and the extensor compartment of the leg (deep peroneal nerve) and extensor digitorum brevis.

Sensory – Upper portion of lateral leg and branch to sural nerve in posterior calf.

Sacral nerves

Some nerves arise directly off the roots before they combine to form the plexus.

Posterior cutaneous nerve to thigh

Sensory – supplies the back of the thigh and upper calf.

Examination of the lower limb summary

Examine the lower limb in general
    shape
    size
    position
    proportion
    alignment.
Examine the stance of the individual.
Examine the gait of the individual
    rate
    rhythm
    cadence
    limb–antalgic, short leg, neurologic.
The Trendelenburg test.

Examine local pathology in the lower limb.

Look

Examine the skin
    colour changes
    inflammation
    bruising

    lacerations
    callosities
    swellings
    effusions
    muscle wasting.
Look for deformities
    bone
    joint
    foot.

Feel

Feel the contours of the bones and joints.
Feel any lumps or swellings
    size
    shape
    consistency
    edge
    lobulation
    fixation
    fluctuation
    pulsation
    transillumination.
Feel the pulses.
Feel for the following:
    temperature differences
    tenderness
    gaps
    bony irregularities
    joint irregularities
    swellings
    joint line tenderness
    synovial thickening
    effusions
    crepitus
    bony bossing.

Measure

Measure for muscle wasting.
Measure
    angular deformities
    rotatory deformities
    leg length discrepancies – femur or tibia

Move

Determine the active and passive range of movements
  hip – flexion, adduction, abduction, rotation
  knee – flexion, extension
  ankle – plantarflexion, dorsiflexion
  subtalar – varus, valgus
  midtarsal – internal rotation, external rotation.
Assess for extensor lag.
Assess for contractures.
Assess for tendon ruptures
  quadriceps
  tendo achilles.
Assess the stability of joints
  hip
  knee – collateral ligaments, cruciate
  ligaments
  ankle – collateral ligaments.
Test for meniscal damage.

Neurological examination

Assess the relevant branches of the lumbar and sciatic plexes.
Lumbar plexus
  Lateral cutaneous nerve of thigh
  Femoral nerve
  Obturator nerve.
Sacral plexus
  Sacral nerves
    Posterior cutaneous nerve to thigh.
  Anterior divisions
    Nerve to obturator internus
    Nerve to quadratus femoris
    Medial popliteal nerve (via sciatic).
  Posterior divisions
    Sciatic nerve.

6 Laboratory investigations relevant to orthopaedic conditions

(Result ranges are given for adults. N.B. Establish the values for your own area as each laboratory has its own levels and ranges of normal. There may be variation from the values given here.)

Haematological investigations

Haemoglobin (Hb) levels
Sample: Whole blood (EDTA).
Test: Haemoglobin can be converted to cyanmethemoglobin and the colour assessed electronically. At present the assessment is automated and directly measured after haemolysis of the blood.
Normal levels: males: 15.5 ± 2.5 g/dl
females: 14.0 ± 2.5 g/dl
Clinical: Haemoglobin is required for oxygen transport in the body.
Low haemoglobin levels are found in the chronic anaemias, acute blood loss following trauma or surgery, diminished quantity of myeloid tissue, as in the myeloproliferative disorders or osteopetrosis (where bone has replaced the marrow).

Haemoglobin electrophoresis
Sample: Whole blood (EDTA).
Test: Haemolysate from whole blood. Electrophoresis on cellulose acetate.
Clinical: The test indicates abnormal haemoglobins A, S and F. Sickle cell disease, thalassaemia major and related conditions can result in pain from infarctions in bone and joints.

Erythrocyte sedimentation rate (ESR)
Sample: Whole blood (Na citrate).
Test: Westergren – centimetres of sedimentation of blood in a tube in 1 hour at 20°C.
Normal range: adult male: 17–50 1–7 mm/hour
>50   2–10 mm/hour
adult female: 17–50   3–9 mm/hour
>50   5–15 mm/hour
Clinical: The ESR is an indicator to the general health of the individual. It is raised in acute or chronic infections, neoplastic and inflammatory diseases but the measurement is affected by anaemia and the administration of dextran and fat emulsions.

White cell count
Sample: Whole blood (EDTA).
Test: Counted electronically.
Normal range : $4–11 \times 10^9$/litre
Clinical: The white cell count is raised in infective conditions and in leukaemias. May be raised following surgery or trauma.

Tests for disorders of coagulation

Assays of plasma factors
Sample: Plasma – fill a citrate tube with whole blood. Keep it cool and send it to the laboratory immediately for fractionation as these factors, particularly Factor VIII are very labile.
Test: Clotting time of test sample, compared with a control, using a standard solution of plasma clotting factor.
Normal range: 50–150%
Clinical: Specific assays of factors VIII and IX are indicated in cases of haemophilia or Christmas disease, relatively common causes of haemarthroses and intramuscular bleeds.

Bleeding time
Test: The time taken for bleeding to cease from a small puncture wound. The test is standardised by using a template which produces an incision of known width and depth. The test is performed with a sphygmomanometer cuff around the arm inflated to 40 mm mercury.
Normally: 2.5–9.5 minutes.
Clinical: The bleeding time is prolonged in cases where there is capillary fragility, or platelet dysfunction. It is affected by aspirin therapy.

Clotting time
Sample: Whole blood in plain glass tube.
Test: The time whole blood takes to clot in a test tube.

Normally: 5–11 minutes.

Clinical: Clotting time is prolonged with clotting disorders. This test is performed in the laboratory under controlled conditions and at 37°C. In the ward situation it is easy to remove a millilitre of blood into a standard glass tube. Warm this tube in your hand, without undue shaking, and time it until the blood clots.

Prothrombin time (PTT), prothrombin index (PI), and international normalised ratio (INR)

Sample: Whole blood in sodium citrate (9:1 ratio critical).

Test: The time taken for citrated plasma to clot when mixed with tissue factor (brain extract) and calcium chloride. The prothrombin index is the ratio of the time the test sample takes to clot against a control of normal blood.

The INR uses a standardised sample of brain tissue as tissue factor and the test is internationally comparable.

Normal range: PTT 24–34 seconds. (depends on thromboplastin)

    PI 12–17

    INR < 1.2 (therapeutic range 2–4)

Clinical: These tests indicate levels of prothrombin. Low levels are associated with liver disease, fibrinogen deficiency e.g. after intravascular coagulation, and vitamin K deficiency, and anticoagulant drugs. Low levels are also prolonged in some hereditary deficiencies of factors V and X. The test assesses efficacy of anticoagulant therapy.

Fibrin degradation products (XDPs)

Sample: Citrated, EDTA or heparinised blood.

Test: Agglutination using polystyrene particles coated with monoclonal antibodies to the human fibrin degradation product D dimer.

Normal: less than 200 ng/ml

Clinical: The plasma of patients with various thrombotic disorders contains crosslinked fibrin derivatives (XDP). (Note that FDPs are formed by degradation of fibrin and fibrinogen.)

Acute defibrination is caused by intravascular clotting resulting from tissue factor entering or being released in the blood stream. Fibrin is laid down on the walls of blood vessels by the flow of blood, rather than forming large thrombi.

This condition can occur as a complication of major trauma, incompatible blood transfusions, pulmonary embolism, deep venous thrombosis and disseminated intravascular coagulation arising from whatever cause. Fibrinolysis of the deposited fibrin occurs as more fibrin is deposited. The result is a depletion of fibrinogen in the blood and this leads to a clotting deficiency. The resultant bleeding is usually massive and is difficult to control.

Biochemical tests

Acid phosphatase

Sample: Serum. Analyse immediately or use citrate or acetate buffer. (Red blood cells contain an acid phosphatase and care must be taken to prevent haemolysis.)

Test: 4-nitrophenol-phosphate or immunochemical analysis.

Normal range: 3.3–5.9 units/litre

Clinical: Normal serum contains only small amounts of acid phosphatase, which arises mainly from the liver and spleen. Prostatic tissue produces large quantities of this enzyme which normally is passed into the prostatic secretions. Prostatic carcinoma, particularly if metastatic, may cause marked elevation of acid phosphatase levels.

Alkaline phosphatase

Sample: Serum. collect in plain tube. (Anticoagulants cause inhibition.)

Test: Automated

Normal range: 73–207 units/litre

Clinical: Serum alkaline phosphatase in normal serum originates almost exclusively from the bones, where it is formed in large amounts in actively functioning osteoblasts. It can also originate in liver, intestine, placenta and lung, and serum levels can be raised by pathology in these organs.

The bone isoenzymes in serum are elevated in cases where there is osteoblastic proliferation or increased activity. Elevated serum levels generally imply one of three situations:

- Inadequate mineralisation (rickets, osteomalacia, hypophosphatasia)
- Injury where the matrix is normal (bone fracture, ectopic calcifications)
- Extensive bone disease (osteoblastic metastatic carcinoma, Gaucher's disease, the severe form of osteogenesis imperfecta and Paget's disease).

Arterial oxygen partial pressure (p0$_2$ level)

Sample: Arterial blood collected anaerobically in a syringe previously flushed with heparin.

Test: Automated analysis with electrodes.

Normal range: 12–13 kPa

Clinical: The partial pressure is lowered in conditions reducing the oxygenation of the blood. Occurs with respiratory depression and pulmonary pathology. It is especially relevant in the adult respiratory distress, particularly the 'shock' lung syndrome after trauma and associated with severe infections and septicaemia.

Calcitonin

Sample: Plain glass evacuated tube.

Test: Radio-immune assay using goat anti-calcitonin serum.

Normal range: 61 ± 20 picogram/ml

Clinical: Calcitonin is secreted by the parafollicular cells in the interstitial tissue between the follicles of the thyroid gland. It is normally present in very low quantities, but levels may be raised in cases of medullary thyroid carcinoma and carcinoma of the lung. Calcitonin acts very rapidly to reduce the level of ionised calcium in the plasma. It acts by inhibiting the absorption of bone, presumably by acting on the osteoclasts.

Therapeutically it is administered, generally as salmon or porcine extract, to reduce the bone activity in Paget's disease.

Calcium (total)

Sample: Serum. Collect fasting and with minimal venous occlusion.
Test: Spectrophotometer
Normal range: 2.1–2.5 mmol/litre
Clinical: Calcium ions are important in the formation of the skeleton; in membrane stability, particularly of nerve and muscle cells; and in blood coagulation. The serum calcium level is dependent on:
(i) the absorption of this element from the upper bowel, influenced largely by vitamin D (The renal excretion is relatively constant and does not play an important, immediate, regulatory role in serum calcium levels.)
(ii) its deposition in bone, governed by calcitonin, oestrogen and other anabolic hormones
(iii) mobilisation from the bones, governed largely by parathormone and vitamin D.
Excessive serum calcium levels are the result of adenomas or carcinomas of the parathyroid glands, excessive vitamin D levels, metaplastic neoplasms of bone, enforced immobilisation, hyperphosphatasia and the 'milk-alkali' syndrome.
   Prolonged hypercalcaemia gives rise to nausea, depression and abdominal pain; while on X-ray, osteoporosis, bone cysts and erosions may be seen. Primary hypercalcaemia may also be associated with renal calculi.
   Lowered levels of serum calcium result from hypoparathryoidism, steatorrhoea, nephrosis and nephritis (leading to secondary hyperparathyroidism) and acute pancreatitis. Hypocalcaemia, clinically, gives rise to tetany.

Calcium (ionised) (Cal)

Sample: Collect anaerobically in a heparinised vacuum tube, without using a tourniquet on the arm. Place on ice.
Test: Automated test using electrode calibrated against a standardised calcium solution.
Normal values: 1.12–1.32 mmol/litre.
Clinical: Assess physiologically active or free calcium and better reflects calcium metabolism. Serum ionised calcium levels are controlled by the parathyroid glands and affected by vitamin D. Raised levels are found in primary hyperparathyroidism, PTH producing tumours, excess intake of vitamin D and various malignancies (especially Ca lung).

Parathyroid hormone (PTH)
Sample: Plain, EDTA or heparinised tubes. Collect in morning after overnight fast.
Test: radio-immune assay using labelled antibodies to PTH.
Normal value: 10–70 picogram/ml
Clinical: PTH is a small protein molecule, secreted mainly by the chief cells of the parathyroid gland. The effect of this hormone is immediately to raise the level of circulating ionised calcium in the extracellular fluids and to reduce the level of circulating phosphate. It does this by its effects on:
● Bone – by directly affecting the osteoclasts it stimulates absorption of bone and the release of calcium into the plasma.
● Kidney – it acts on the tubules of the kidneys to cause an increase in the reabsorption of calcium and a rapid loss of phosphate in the urine.
● Gut – it stimulates increased intestinal absorption of calcium provided vitamin D is available.
Excess secretion in primary or secondary hyperparathyroidism leads to hypercalcaemia, with its attendant symptoms and to osteoporosis and bone cysts.
Lowered PTH levels are associated with hypocalcaemia and a normal phosphate level. Clinically this condition arises idiopathically or follows surgery to the thyroid gland and manifests with tetany.

Phosphate
Sample: Serum – plain tube.
Test: Spectrophotometer.
Normal range: 0.56–1.30 mmol/litre
Clinical: Phosphate is an abundant compound and abnormal levels rarely cause a clinical problem. It plays an important role in organic metabolism, particularly in energy storage and transfer by way of high-energy phosphate bonds. The serum level is subject a circadian rhythm. There is a rough reciprocal relationship between serum calcium and phosphate levels: an elevated phosphate may cause a fall in the serum calcium.
Levels are raised in osteolytic metastatic bone tumours, myelogenous leukaemia, sarcoidosis, milk-alkali syndrome, vitamin-D intoxication, renal failure and hypoparathyroidism. They are lowered with osteomalacia, hyperparathyroidism, steatorrhoea, renal tubular acidosis, acute alcoholism and gram negative septicaemia.

Proteins and protein electrophoresis
Sample: Serum.
Test: Automated. Electrophoresis on cellulose acetate.
Normal range: 60–80 g/litre
Clinical: Serum proteins are a mixture of albumin, and three globulin fractions – alpha, beta and gamma. Levels are raised in polyclonal or monoclonal gammopathies (demonstrated with electrophoresis). They are lowered with starvation, burns, gastroenteropathies and nephrotic syndrome.

Urate

Sample: Serum.
Test: Phosphotungstate.
Normal range: 0.20–0.42 mmol/litre
Clinical: Urate reflects purine breakdown within the body and is normally excreted in the urine. Raised blood levels may be due to:

- Decreased excretion – as occurs in renal disease.
- Haematological disorders – raised levels indicate a high metabolic turnover and is seen in the leukaemias, lymphomas, multiple myeloma, and polycythaemia.
- Gout – an inherited condition characterised by a derangement of purine metabolism. It is thought that there is an excessive rate of conversion of glycine to the purines, which are then degraded to uric acid.

   Although hyperuricaemia occurs with the acute episodes, elevated levels of serum urate , *per se*, do not give rise to the attacks of gout. It is the precipitation of uric acid crystals within the tissues, in Gout, which is responsible for the symptoms, and does so by stimulating an acute inflammatory response in these areas.

Vitamin C

Sample: Plasma (oxalate or heparin).
Test: Colorimetrically – dinitrophenylhydrazine.
Normal range: 25–85 micromol/litre
Clinical: Vitamin C is essential for the maintenance of intercellular substances. These include the fibrils and collagen of connective tissue.

   Vitamin C levels are lowered with nutritional lack (scurvy), haemodialysis, anaemia, pregnancy, alcoholism, malabsorption and hyperthyroidism. The pathologic signs are confined to the supporting tissues of mesenchymal origin and manifest as subperiosteal haemorrhages, loosening of teeth, poor wound healing and fractures of bones.

Vitamin D

Sample: Serum
Normal values: Vitamin D (24,25 dihydroxy) 8.4 ± 3.4 nmol/litre
   Vitamin D2 (25hydroxy) 9.4 ± 7.5 nmol/litre
   Vitamin D3 (1,25 dihydroxy) 100 ± 28 pmol/litre
Clinical: The effect of vitamin D is to increase the availability and utilisation of calcium and phosphorus in the plasma by increasing intestinal absorption and increased bone resorption. It governs the long term regulation of the serum ionised calcium level.

   Vitamin D is obtained from the diet or through the action of sunlight on precursors in the skin. For activity, activation of the vitamin, by hydroxylation, takes place in the liver and then in the kidneys.

   Raised levels of vitamin D are well tolerated and only cause symptoms of hypercalcaemia when present in extremely large amounts (500 to 1000 times the normal requirements). Moderately raised levels of Vitamin D3, the most active form, occur in tumoral calcinosis, primary hyperparathyroidism, sarcoidosis, growing children and lactating women.

The clinical importance of vitamin D is seen when it is deficient. A lowered level of vitamin D occurs with deficient exposure to sunshine, hypoparathyroidism, chronic renal failure and anephric patients. The serum phosphate usually falls while the level of serum calcium usually remains normal, except in infants.

The net result is deficient mineralisation of osteoid tissue in children, giving rise to rickets; and diminished mineral content in the bones in adults, resulting in osteomalacia.

Serological tests

Antistreptolysin O (ASO) Titre
Sample: Serum.
Test: Serological.
Normally: Normally negative. May be raised after exposure.
Clinical: Streptococcus A antigens can be responsible for rheumatic fever, including its joint manifestations, and acute glomerular nephritis in susceptible people.

Brucella
Sample: Serum.
Test: Agglutination.
Normal: Negative (low titre in exposed individuals)
Clinical: Brucellosis, mainly a disease of cattle and goats, can be responsible for vertebral osteomyelitis or a monarticular septic arthritis.

Echinococcus
Sample: Serum
Test: Serological
Normally: Negative.
Clinical: Echinococcus is a tape worm which infests dogs. It isssentially an infection transmitted through sheep but the larvae of the echinococcus sp. can gain access to the human. Hydatid cysts then result in the liver, lung, and bone.

Fungal organisms
Sample: Serum.
Test: Serological.
Normally: Negative.
Clinical: Fungi are opportunistic pathogens, infections occurring when there is a lowered immunity to infection on the part of the patient, resulting from malnutrition, infection or other deficiency of the immune system. A variety

of fungal organisms including aspergillosis, blastomycosis and cryptococcus, are occasionally responsible for a chronic osteomyelitis. Apart from media culture of the organism, specific serological tests are available.

Rheumatoid factor
Sample: Fresh serum. Plain tube.
Test: Agglutination of latex particles coated with human IgG.
Normally: Negative.
Clinical: Rheumetoid factor is found in many patients with rheumatoid arthritis, and in other collagenoses. Note that cases of rheumatoid arthritis do occur in which the rheumatoid test is negative – the so-called 'seronegative' rheumatoid arthritis.

Syphilis
Sample: Serum
Test: VDRL (venereal disease research laboratory) or Reagin.
Normally: Negative
Clinical: Luetic disease is still prevalent in some parts of the world. Classically, bone lesions affect the face and long bones of infected neonates. Involvement of the spinal cord affects the long posterior tracts and gives rise to tabes dorsalis. In these cases there is diminished pain and proprioceptive sensation to the legs. This characteristically results in a broad stepping gait and 'Charcot' joints, especially of the knees.

Yersinnia
Sample: Serum.
Test: Agglutination.
Normally: Negative.
Clinical: An enteric pathogen, this organism is occasionally responsible for an acute, monarticular arthritis.

Skin tests

Hydatid disease
Method: Cassoni test – injected intradermally.
Test: Sterilised fluid from sheep hydatid cyst.
Clinical: Hydatid disease can give rise to fluid filled cysts in bone which contain brood capsules of the eccinococcus. A large weal, surrounded by erythema, indicates present or past infestation with the organism.

Tuberculosis
Method: Tuberculin, Mantoux and Heaf tests – injected intradermally.
Test: Tuberculin, concentrated from the filtered supernant of a culture of Mycobacterium tuberculosis.

Clinical: Tuberculosis of bone and joint still exists in a significant number of cases. A florid reaction to the skin test suggests active infection. These tests are also valuable indicators of previous exposure to tuberculosis and the consequent resistance or susceptibility to infection.

Urine tests
Excess or abnormal substances can be found in the urine in a number of pathological conditions related to orthopaedics.

Amino acids
Sample: Urine random.
Test: Column chromatography.
Normal range: 3.6– 4.3 mmol/dl
Clinical: Urinary amino acid excretion is diminished with protein malnutrition. Urinary amino acid excretion is increased in liver failure, acute renal tubular necrosis, Fanconi's syndrome, Wilson's disease and aminoaciduria.

Bence jones proteins
Sample: Random urine
Test: Previously heat precipitation, now by electrophoresis
Normal range: Nil
Clinical: These proteins form a heterogenous group of low molecular weight proteins that occur in association with many cases of multiple myeloma. They are not found in the serum as they are readily diffusible through the glomerular membrane and rapidly filtered out in the urine. They are characterised by precipitating from solution at 40–60°C and redisolving between 95 and 100°C.

Calcium
Sample: Urine (24 hour).
Test: Spectrophotometer.
Normal: 2.5–7.5 mmol/dl
Clinical: Excessive loss is associated with hyperparathyroidism, hyperthyroidism, hypervitaminosis D, nephrosis, acute nephritis, malignant metastases in bone and steatorrhoea, and idiopathic hypercalciuria.
Diminished loss is associated with hypoparathryoidism, rickets and osteomalacia, and hypothyroidism.

Hydroxyproline
Sample: Urine 24 hour
Test: Colorimetric
Normal range: 0.11–0.33 mmol/dl
Clinical: Urinary hydroxyproline is raised in Paget's disease, acromegaly, rickets and osteomalacia, fractures, osteomyelitis and neoplasia.

Mucopolysaccharides

Sample: Urine (fresh random).

Test: Trimethyl ammonium bromide or Toluidine blue. (Toluidine blue is used to stain urine samples on filter paper. The paper is decolourised and dried. Mucopolysaccharides show up as blue rings.)

Normally: Negative.

Clinical: Acid mucopolysaccharides are found in the mucopolysaccharidoses, a number of the congenital syndromes characterised by dwarfism, visceral abnormalities and lysosomal storage of mucopolysaccharides. Hurler's and Morquio's diseases are examples of these conditions.

7 Imaging investigations

Plain radiographs

Anteroposterior and lateral radiographic projections give a negative image of a three dimensional structure (the body), which is projected onto a flat sheet (the radiographic). Bone shows up as white, clear areas and the soft tissues as darker tones. The different densities on the film signify the different absorptions of the X-rays as they pass through the tissues of the body. The greater the absorption of X-rays by the tissues the clearer the image on the plate. The less the absorption of X-rays, the darker the image.

Because of this projection, the structures and their pathological changes shown in each X-ray image need to be interpreted. In order to make this assessment, knowledge of the normal radiographic appearance is essential.

In considering any radiograph, note the following:

Soft tissues

Assess the soft tissues for their shape, their tissue plains, and any swellings   Enlargements, increased densities and obliteration of tissue plains can indicate obesity or swellings from oedema and tumours.

Assess the soft tissues for the presence of gas   Gas in the soft tissues is always pathological and arises from one of three causes – an injury that lacerates the skin; a pneumothorax; or an infection with gas forming organisms.

Assess the soft tissues for radio-opacities   Localised radio-opacities can arise from natural calcifications (atheromatous arteries, phleboliths, calculi, myositis ossificans, tumoral calcinosis) or from foreign material within (or without) the body.

Bones

Assess the bones for their form   Each bone generally has a particular shape. Inherited conditions, growth deficiencies, and pathological disease processes, such as infection, tumour and bone softening can result in a deformity.

Assess the bones for their radiographic density   Diminished bone density is seen when there is reduced mineralisation of the bone, which may be generalised or localised. Generalised increase in bone density is seen in 'marble bone disease'. Localised 'sclerotic' areas indicate excess mechanical stress. Dead bone often appears more dense than the surrounding bone.

Assess the bones for their internal structure   The interior of bone consists of trabeculae. Paget's disease and sclerosing metastatic tumours can cause thickening of these bony struts. Disturbance of the normal trabecular pattern may occur in healing fractures (woven bone), in dysplastic bone, or when cysts or tumours destroy the bone.

Assess the bones for their integrity   Breaks in the cortical margin or in the trabecular lines generally indicate a fracture of that bone.

Joints

Assess the 'joint space'   The radiolucent line between the bony condyles of the joint represents the radiolucent articular cartilage. Cartilage loss manifests as a narrowing of this radiological 'space'.

Assess the joints for the shape of the condyles   The bone metaphysis flares out to form smooth, rounded, articulating condyles of the joint. Inflammatory joint lesions can cause enlargement and squaring of these condyles. Marginal osteophytes will disturb the normally smooth outline of this portion of the bone.

Assess the subchondral bone of the joint margins   The subchondral bone becomes denser in cases of osteoarthritis. Irregularities in this area may arise from fractures, erosions, cysts or tumours

Assess the congruity of their articular surfaces   The apposing bony surfaces should be parallel over the whole of the joint margin. Loss of congruity occurs with destructive lesions of the joint, with subluxations and in dislocations.

Assess the relative positions of the bones to one another   The bones on either side of a joint are usually seen to be in a 'normal', anatomical position. Mal-alignment may indicate an instability or contracture of that joint. Note the alignment of the vertebrae on spinal radiographs.

Certain conditions show distinct radiographic appearances:

Osteoporosis – shows up as a reduction and thinning of the trabeculae, and a thinning of the cortices, giving the bone a washed out appearance. These changes are more easily assessed in the metaphyseal regions of the long bones and in the vertebrae; although thinning of the cortex and widening of the medullary cavity are also seen in the diaphyseal regions.

Osteoarthritis – idiopathic or associated with other pathologies in the joint, is seen radiologically as a reduction in the width of the 'cartilagenous' joint space; accompanied by projecting 'osteophytes' along the joint margin. There are also commonly sclerotic thickening and radiolucent 'cysts' in the subchondral bone on either side of the joint.

Bone tumours – generally show up as irregular lucencies in the bones. A tumour generally destroys the surrounding bone but in some cases it may 'expand' the bone. Cartilagenous tumours may show irregular areas of calcification within them.

A thickened, sclerotic, continuous bony margin around the lucency suggests a relatively benign lesion. A thin, irregular, eroded bony margin suggests a more aggressive tumour which has infiltrated into the surrounding bone. In malignant tumours a diffuse infiltration into the surrounding bone may be noted.

Bone softening – (rickets, Paget's disease, osteogenesis imperfecta) is indicated by bowing of the weight bearing bones.

Fractures – show up as breaks in the smooth continuity of the cortex and the internal structure of bone. The bony fragments may have separated, often with mal-alignment (linear, angular or rotatory). Some fractures are difficult to visualise on the X-ray. In these cases their presence must be deduced from displacements of the normal relationships of the bones.

Stress views of a region may show up abnormal relative movement of two bones or fragments. This is particularly useful in determining the stability of joints or fractures, such as the ankle and in the spine.

Tomography

Tomography shows structures radiographically in a pre-selected plane by blurring out structures above and below this plane by rotating the X-ray tube and plate around this selected plane. This technique is an adjunct to plain films and is useful when radiographic detail is obscured by overlying structures on the initial plates.

Contrast studies

Radiographic visualisation of a pathological process, particularly those involving soft tissues, may be enhanced using contrast material that has a greater absorption of the X-rays. The contrast medium, generally in a cavity or channel of some form, will outline the area requiring investigation. These studies may be performed using only contrast material (single contrast) or further enhanced by using air and contrast material (double contrast). Commonly these studies are utilised in the following situations:

Sinography   – to demonstrate the tract of a discharging sinus.

Myelography – demonstrates the soft tissue structures around the spinal cord.

Discography – demonstrates the structure of the intervertebral disc.

Arthrography – demonstrates soft tissue pathology in synovial joints.

Angiography – demonstrates aneurysms, vascular trauma, and tumours.

Venography   – demonstrates the patency of veins and other venous channels.

Computerised axial tomography (CAT) scans

Cross sectional X-rays, now taken by a fan of X-ray beams, are integrated by computer and projected to give axial 'slices' through the area under investigation. Bone shows up as pale, radiodense areas. The different soft tissues can be assessed and identified by measuring their radiographic density.

This technique is particularly useful in assessing bony pathology. The image can be reformatted in different planes, including three-dimensional visualisation. Measurements of distances and angles can also be performed.

Magnetic resonance imaging

MRI scans are produced by measuring and computer compilation of radio signals induced in certain atomic nuclei sited in a magnetic field and acted on by a suitable radio frequency pulse.

Of the nuclei that respond in this fashion, hydrogen is the most abundant in the body. Accordingly those tissues with a high water content show up best. In the scan these areas show up white. Bone, having relatively low quantities of hydrogen, shows up black.

T1 and T2 weighting give slightly different imaging characteristics to the different tissues. These techniques are particularly useful in soft tissue lesions such as ligament injuries and tumours affecting bone.

Photon absorptiometry

The relative absorption of gamma radiation from a radioactive source by bone is measured. The quantity absorbed will give a measure of the degree of calcification of the bone and can be compared with a standard.

Sonography

Ultrasonic scans are useful in assessing soft tissue lesions, particularly around joints. As there is no radiation and they are non-invasive, these scans are often used to assess congenital dislocations of the hip in neonates.

Bone scans

Radio-isotopes are useful diagnostic tools for orthopaedic problems, particularly in cases of fracture, osteoarthritis, infection, and tumour. In some cases, areas of bone necrosis can be shown up by the diminished uptake of isotope in the affected area.

Technitium MDP (planar) scans

Diphosphonate, labelled with radioactive technitium shows up increased metabolic activity in bone and can demonstrate a variety of conditions. These include bone tumours, infection, fractures, arthritis, avascular necrosis, metabolic bone disease and Paget's disease. Clinically the scan may be positive in two phases: an immediate angiographic phase, when the radionuclide is in the blood stream (dynamic scan), and a later 'delayed phase' when the labelled diphosphonate is thought to be taken up by areas of new bone formation (static scan).

SPECT (single photon emission computerised tomogram) scans

Several images are taken radially around the area of investigation and then reconstructed by computer, as with radiographic CAT scans. This is a useful technique where the plain scan is equivocal.

Gallium scans

Radioactive gallium citrate is a useful radionuclide which is taken up by bone tumours and infective lesions of bone.

Indium labelled leukocyte scans

Leukocytes are separated from the patient's serum. These are incubated with radioactive indium and then reinjected into the patient. They congregate at the site of an inflammatory process and are indicated by the standard scanning camera. This scan can show up areas of acute and chronic bone infections.

8 Orthopaedic trauma

Trauma is an important part of an orthopaedic surgeon's practice. Although the most obvious 'orthopaedic' injury may be the fracture of a bone, this may not be the most important injury in that particular patient. In managing 'orthopaedic' injuries it is necessary to have an understanding of the effects of injury on the body, both locally and generally.

Local effects of trauma

Trauma has direct effects on bones and on soft tissues.

Bones

Bone has an innate strength and elasticity and can absorb most of the forces to which it is ordinarily subjected; but it will break if the force applied is greater than it can contain. A fracture, therefore, represents a failure of that bone to contain and absorb the forces to which it has been subjected. As such it is a relative event and depends on the strength of the individual bone and the amount of force applied.

In the development of a fracture the rate of loading of the bone is important. A strong force applied relatively slowly will accumulate along a single line, which then fails. Rapidly applied force does not have time to concentrate and the lines of failure spread out in many directions. The bone literally 'explodes', giving the comminuted fracture.

Local signs of a fracture

Clinically the following signs, together, are pathognomonic for a fracture:
1. Tenderness
2. Deformity
3. Swelling
4. Loss of function
5. Bony crepitus.

Soft tissues

Although the bone is the structure most resistant to bending and, as such, it bears the initial force that is applied, it does not follow that the applied force ceases after the bone breaks. Indeed the force may continue unabated, in which case the soft tissues now take its brunt, twisting and tearing, causing damage to the skin, muscles, fascia, nerves and vessels.

The blood supply to a limb may be impaired in one of three ways.

1. The artery may be damaged by the injuring force directly or by the bone fragments as they are displaced. Severance of the artery will be shown by absence or diminution of the distal pulses. (Occasionally only the intima of the artery is injured. In these cases the distal pulses initially appear relatively normal. It is only a few hours later that they disappear as thrombus occludes the vessel.)
2. Damage to the small vessels of the limb leads to sludging of the blood within them and progressive intravascular thrombosis. Tissues which initially appear viable are found to be necrotic a day or two later.
3. The small vessels of the limb may be occluded by progressively increasing pressure within the compartment of the limb as it swells due to the accumulation of oedema fluid (compartment syndrome).

All these mechanisms lead to an acute loss of the blood supply to the tissues of the limb and its inevitable necrosis.

Paradoxically, when the small vessels only are affected, the deep tissues may be necrotic while the overlying skin appears relatively normal. Important, albeit late, signs are the loss of function of the muscles and loss of cutaneous sensation.

Important facets of examination are the assessments of the distal pulses and of the tissue pressure of the injured limb.

Impaired blood supply to the bone greatly impairs its healing and leads to an increased risk of infection. This is particularly important in 'compound' fractures where the skin overlying the fracture is torn, offering an opportunity for bacterial colonisation of the injured area.

General effects of trauma

Forces which fracture bones may also directly injure other organs, the brain, the lungs, the heart, and the abdominal viscera. A complete examination of these structures is necessary in the polytraumatised individual.

Severe trauma on bones may lead to widespread indirect effects on the individual, producing damage in other structures. Hypovolaemia, circulating endotoxins, products of tissue breakdown and the metabolic changes of major trauma act detrimentally on the heart, lungs, kidneys, liver and brain.

Irreversible hypovolaemic shock

A fracture is always associated with some quantity of blood loss. In a compound fracture, with an overlying skin laceration, this blood loss is obvious. The blood loss associated with a closed fracture is not so obvious and varies from a few millilitres with a hand fracture; to a 'unit' with a fracture of the tibia; to two 'units' with a fracture of a femur; to up to 15 litres with some fractures of the pelvis. The hidden blood loss with multiple fractures is cumulative and must be considered in the assessment of the traumatised patient or risk his sudden collapse.

Vascular hypotension can lead to the release of endotoxins from the bowel and to renal failure. These factors will complicate the management of any patient. Hypotension due to blood loss, if prolonged, will go on to a hypovolaemic shock which is irreversible.

Septic shock

Polytraumatised patients are susceptible to infection, arising particularly in their lungs and in their wounds. Bacterial septicaemia is a common complication of the polytraumatised patient and leads to multiple organ failure, including that of the heart, liver and kidneys, and consequently to the ultimate demise of the patient.

'Shock' lung

A variety of factors, including chest injury, hypovolaemia and marrow embolisation from fractures of the long bones have been incriminated in the pathogenesis of the 'shock' lung syndrome. Prealveolar arteriovenous shunting occurs in the lungs, giving a ventilation/perfusion imbalance. This results in generalised tissue anoxia and ultimately in gross metabolic damage to the brain, lungs, liver, kidneys and heart.

(Early operative stabilisation of fractures has been shown to reduce the incidence of these complications in these severely injured patients.)

'Crush' syndrome

Prolonged ischaemia of a limb, from its position or vascular injury, will lead to the 'crush' syndrome when the vascular circulation to that area is restored. The sudden influx of metabolic substances and products of tissue breakdown into the circulation go on to block the renal tubules and result in acute renal failure.

Cerebral death

A prolonged period of anoxia to the patient, from hypovolaemia, his position, airway secretions or chest injury can lead to severe brain damage or death.

Thus

In the general management of the acutely injured person it is essential to:
1. Ensure his airway is clear
2. Restore the circulating blood volume
3. Repair injured vessels
4. Stabilise fractures

Recommended Reading

For further reading on this subject the following, more extensive, texts are recommended:

Apley's System of Orthopaedics and Fractures
Apley AG and Solomon L
Butterworth/Heinemann, 7th Edition, 1993

Clinical Orthopaedic Examination
Ronald McRae
Churchill Livingstone, 1990

Orthopaedic Diagnosis and Management
Boyd S Goldie
Blackwell Scientific, 1992

Orthopaedic Physical Assessment
David Magee
WE Saunders, 1987

Outline of Orthopaedics
J Crawford Adams
E & S Livingstone

Seven Paths to Orthopaedics
TL Sarkin
Maskew Millar Longman, 1989

Index

Abcess 2, 3, 21, 22
Abduction 44, 48–51, 54, 64, 80
Abduction contracture 78, 82
Ability to lie flat 28
Abnormal haemoglobins 92
Abnormal temperature 73
Achondroplasia 9
Acid phosphatase 95
Acquired deformities 4, 46
Acromegaly 17
Acromion 44, 52
Active movements 49
Acute blood loss 92
Acute urinary retention 9
Adduction 64, 78, 80
Adduction contracture 78, 81
Adult respiratory distress syndrome 15, 95
Alcohol 9, 10, 16
Alcoholism 10, 14, 41, 97, 98
Alignment legs 65
Alimentary system 15, 16
Alkaline phosphatase 95
Amelia 41
Aminoaciduria 101
Amputation 41, 65
Anaemia 10, 93, 98
Anaesthesia 8, 14, 15, 85
Analgesics 3, 9, 10
Aneurysm 22, 30, 106
Angiography 106
Ankylosing Spondylitis 28
Ankylosis 23
Antalgic limp 66
Antistreptolysin O Titre 99
Apert's syndrome 46
Appendicitis 34
Arachnoidactally 46
Arches of foot 64, 69, 70
Ariboflavinurix 14
Arm 40, 43, 51, 57, 59
Arterial embolus 43, 73
Arterial oxygen partial pressure 95

Arterial rupture 43, 73
Arthritic joint 2
Arthrodesis 24
Arthrography 106
Arthroplasties 7
Athetoid movements 18
Autonomic function 38
Autonomic nervous system 21, 38
Avascular necrosis 10, 78, 107

Backache 9, 16, 34
Bence Jones proteins 101
Beriberi 14
Bleeding tendency 12
Bleeding time 93
Blood pressure 8, 15
Blood transfusions 94
Bone density 104
Bone dysplasia 7, 76
Bone scans 107
Bone softening 105
Bone tumours 97, 105, 107
Bones
    clavicle 43, 44, 46, 50
    femur 78, 84, 86, 110
    humerus 43, 51
    metatarsal 72
    radius 44–46
    scapula 41, 46, 51, 57
    tarsal 72
    tibia 18, 65, 72, 77, 78, 85, 86, 90, 110
Bony bossing 21, 74
Boutonierre deformities 46
Bowing 4, 29, 41, 46, 76, 77, 105
Brachial plexus 56
Bronchopneumonia 13
Brucellosis 99
Bruising 19, 40, 67
Bunions 71
Burger's disease 42
Burn 4, 23

Cachexia   14
Cadence   66
Cafe au lait spots   18
Calcaneus   64, 82
Calcitonin   95
Calcium   96, 101
Calculi   9, 16, 96, 103
Calipers   6, 66
Callosities   20, 41, 67, 71
Campodactyly   41
Carcinoma lung   96
Cardiac failure   15
Cardiac myopathy   14
Cardiac thrombosis   15, 73
Cardiovascular system   8, 12, 15, 16
Carrying angle   40
Cassoni test   100
Casts   3, 20, 40
CAT scans   106
Cavus   64, 70
Central nervous system   16, 26, 37
Centre of gravity   66, 67
Cerebral Palsy   4, 16, 18, 48, 70
Cerebrovascular accidents   4, 15, 70
Chest pain   8
Chondromalacia patella   72
Choreiform movements   18
Christmas disease   93
Chronic affliction of joints   6
Chronic obstructive airways   15
Circumference leg   72, 74, 75
Claw hand   56
Clawed toes   20, 71
Clawing   41, 48
Clinical tests
    anterior drawer   85
    apprehension test   72
    Barlow   84
    bowstring   34
    Bragard's   34
    cruciate ligament   85
    FABER   34
    femoral stretch   34, 39
    Froment's   60
    Gaenslen   34
    gap   73
    grinding   87
    heel-shake   87
    intrinsic muscles   60
    Lachman   85

    Lasegue's   34
    McMurray   87
    Ortolani   83
    patellar tap   73
    pivot shift   86
    point test   21
    posterior drawer   85
    posterior sag   85
    recurvatum test   85
    sciatic stretch   34
    straight leg raising   34
    Thomas's test   81
    Trendelenburg test   67
Clonus   37
Clostridia welchii   21
Clotting time   93
Clubbing   42
Cobb's angle   30
Collagenoses   8, 9, 15, 18, 100
Congenital abnormalities   46
Congenital deformities   4, 15, 41
Congenital dislocation hip   83, 84
Contracture   4, 23
    abduction   78, 82
    adduction   78, 81
    Duypuytren's   42
    hamstring   82
    hip   78, 81, 82
    hip flexion   81
    joints   24
    quadriceps   82
    tendo achilles   82
    tensor fascia   82
    upper limbs   48, 49
    Volkman's 4
Coronary thrombosis   15
Crepitus   21, 43, 74, 108
Crohn's disease   7
Cruciate ligaments   85
Crutches   6, 8, 66
Cubitus varus   46
Cyanosis   15, 42
Cyst   21, 22, 100

Dampness of the skin   21
Deep vein thrombosis   8, 14, 17, 94
Deformity
    angular   4
    commencement   2, 4, 6
    duration   4

lower limb   68, 75
nature   3, 4
progression   4, 6
site   3
spine   29
upper limb   41
Degrees of pain   10
Dehydration   14, 15
Delerium   10, 16, 42
Dementia   14
Deprivation   2
Dermatitis   14
Dermatome innervation   37
Diabetes   8, 16, 41, 43, 73
Diagrams
    angulations of knee   76
    anterior drawer   86
    Barlow's test   84
    boutonniere deformity   47
    brachial plexus   57
    bunion   71
    cavus foot   70
    claw hand   42
    clawing of toes   71
    club foot   70
    Cobb's angle   30
    dermatomes   36
    external rotation deformity   68
    finger flexion deficiency   55
    foot rotation   65
    forearm rotation   53
    Froment's sign   60
    hip abduction   80
    hip extension   79
    hip flexion   78
    hip rotation   79
    kyphosis spine   28
    kyphotic angle   32
    leg internal torsion   69
    lordosis spine   28
    lordotic angle   33
    mallet finger   48
    McMurray's test   87
    measurement leg length   77
    nerves in hand   59
    Ortolani's test   84
    patellar test   74
    posterior sag   86
    rheumatoid hand   47
    rib hump   29

scoliosis   29
shoulder abduction   51
shoulder external rotation   52
shoulder flexion   50
shoulder internal rotation   51
spinal alignment   27
swan neck deformity   47
Thomas's test   81
upper limb contracture   49
varus heel   69
wrist dorsiflexion   53
wrist volarflexion   54
Diaphyseal dysplasia   17
Diarrhoea   9, 14
Diphosphonate   107
Discography   106
Dislocations
    acromioclavicular   44
    elbow   44, 46, 52
    foot   72
    hand   45
    shoulder   44
    sterno-clavicular   44
    wrist   45, 46, 53
Dorsiflexion   53, 54
Dowager's hump   29
Drop foot   66
Drug history   9
Dryness of the skin   21
Dupuytren's contractures   42
Dwarfism   17, 65
Dyspnoea   8

Echinococcus   99
Economic status   2
Ectromelia   41
Effusion   22, 23, 43, 68, 73
Ehrlos-Danlos syndrome   18, 83
Epiphyseal dysplasia   17
Equinus   64, 70
Erb-Duchenne paralysis   57, 63
Erythrocyte sedimentation rate   93
Examination
    general orthopaedic   16
    general systematic   13
    local orthopaedic   18
    lower limb   64
    spine   26
    upper limb   40
    neurological   88, 91

Extension   49, 50, 52, 54–56, 62
Extensor lag   80
Extensor tendons   56
External rotation   50, 52, 68, 79

Family history   9
Fasciculation   20, 37
Femoral anteversion   77
Fever   8, 15–18, 99
Fibrin degradation products   94
Fibrinogen   94
Fibrosis   4, 19, 23
Fibrositis   31
Fingers   54
Flacid paralysis   35
Flat foot   70
Flexion   32, 33, 50, 52–55, 78
Flexor tendons   55
Fluid retention   14, 15
Foot   20, 21, 65, 67, 68, 70–72
Foot arches   64, 69, 70
Forearm   41, 46, 51–53, 57, 60
Fracture fixations   7
Fractures   21, 46, 52, 68, 74, 108, 111
    clavicle   43
    hand   44
    humerus   43
    patella   72
    radius   44
    spine   28, 30
    ulna   44
    wrist   44
Frozen shoulder   52
Function   6, 7, 16, 35, 40, 48, 55, 56,
    66, 88
Fungal organisms   99
Fungus   20

Gait   66, 67, 76
Gallium scans   107
Game keeper's thumb   45
Gaps   4, 31, 73
Gastrectomy   7
Gastric erosions   10, 15
Gastro-intestinal problems   9
Gaucher's disease   95
General surgical procedures   7
Generalised laxity of joints   18
Generalised limb or joint disease   17
Generalised orthopaedic problems   7

Genito-urinary system   16, 18
Genu valgum   75
Genu varum   75
Gibbus   30
Gigantism   17, 41
Giving way   6, 11
Glomerular nephritis   99
Goniometer   50
Gonoccal infection   9
Gout   8, 9, 16, 18, 22, 72, 98
    Tophi   18
Grasp   5, 58
Growth plate   4, 5, 46, 76
Gynaecological procedures   7

Haematemesis   15
Haematuria   9, 16
Haemoglobin   92
Haemophilia   9, 19, 22, 74, 93
Hammer toes   20
Hand   5, 40–42, 44, 46, 54, 55, 58, 59
Heaf tests   100
Health   13
Hepatic dysfunction   16
Hiberden's nodes   42
History
    drug   9
    epidemiological   1
    family   9
    general orthopaedic   7
    medical   8
    occupational   7
    orthopaedic   7
    presenting complaints   2
    social   1
    surgical   7
Hurler's   102
Hydatid   99, 100
Hydration   14
Hypercalcaemia   96–98
Hyperparathyroidism   9, 16, 96, 97,
    99, 101
Hypocalcaemia   96, 97
Hypophosphataemia   9, 41, 68
Hypovolaemia   15, 21, 42, 95,
    109–111

Indium labelled leukocyte scan   107
Infection   1, 2, 9, 17, 19, 20, 30, 74,
    93, 107, 109, 110

Inflammation   18, 19, 23, 34, 40, 67
INR   94
Internal rotation   49, 51, 76, 77, 79
International normalised ratio   94
Interspinous ligaments   31
Intimal damage   43, 73
Intravascular coagulation   94
Intrinsic minus hand   56

Jaundice   15
Joint effusions   73
Joint line tenderness   73
Joint movement   35
Joint space   104
Joint swelling   6, 11
Joints
   ankle   68, 72, 73, 82
   carpo-metacarpophalangeal   54
   elbow   40, 44, 46, 52
   hip   67, 72, 76, 81, 83, 84
   interphalangeal   55
   knee   72, 75, 82,85, 86
   metacarpo-phalangeal   46, 54
   radiohumeral   43
   sacro-iliac   31, 34
   shoulder   44, 46, 48, 50, 51–52
   wrist   40, 43, 45, 46, 53

Klumpke's paralysis   57
Knee   82
Koilonychia   42
Kwashiorkor   14
Kyphosis   28, 31, 32

Laboratory tests   92
Lacerations   19, 26, 40, 67
Langer's lines   19
Leg length measurements   78
Length discrepancies   31, 48, 77
Leprosy   41
Ligamentous laxity   18
Ligaments   6, 31, 49, 54, 73, 83, 85, 87
Limp   66, 89
Liver function   10
Locking   6, 11
Lordosis   28, 33, 38, 66, 81
Loss of function   5, 6, 35, 42, 108
Loss of weight   9, 14
Lower motor neurone   35

Lumbar plexus   88
Lumbar spine   27, 33
Lumps   4, 21, 43, 72
Lupus erythematosis   8

Madelung's deformity   46
Magentic resonance imaging   106
Mal-alignment   105, 105
Malabsorption   7, 98
Malaena   16
Malignancy   2, 21
Malleoli   72, 75, 77, 78
Mallet finger   48
Malnutrition   14
Malunion   41, 46, 68
Mantoux   100
Marble bone disease   104
Marfan's syndrome   18, 46, 83
Medical history   8
Meniscal damage   87
Mental confusion   16
Metaphyseal dysplasia   17
Metastatic bone tumours   97
Metatarsal bones   72
Milk-alkali syndrome   96, 97
Monteggia fracture   44
Morquio's disease   102
Motor Neurone Disease   7, 16, 35
Motor power   37
Movements of the lower limb   80
Movements of the upper limb   49
MRI   106
Mucopolysaccharides   17, 102
Muscle imbalance   4, 6
Muscle stiffness   14
Muscle tone   35
Muscle transfers   7
Muscle wasting   20, 41, 45, 67, 74
Muscles
   biceps   48, 58
   brachialis   58
   coracobrachialis   58
   deltoid muscle   58
   gastrocnemius   82, 83
   gluteus maximus   66
   iliacus   88
   interosseous   41, 50
   latissimus dorsi   58
   obturator externus   88
   pectineus   88

popliteus 89
psoas 88
quadriceps 66, 68, 73, 82
sartorius 88
soleus 82
Muscular development 40
Muscular dystrophies 16
Myaesthenia gravis 18
Myelogenous leukaemia 97
Myelography 106
Myeloproliferative disorders 92
Myositis ossicans 103

Neck 28, 32
Neglect 2
Neoplasms 22, 46, 96
Nerve root irritation 34
Nerves
auxillary 58
femoral 88
lateral cutaneous of thigh 88
lateral pectoral 58
medial pectoral 58
median 58, 59
median cutaneous 58
musculocutaneous 58
obturator 88
obturator internus 88
popliteal 88
posterior cutaneous of thigh 89
quadratus femoris 88
radial 58, 60
saphenous 88
sciatic 89
suprascapular 58
to latissimus dorsi 58
to rhomboids 57
to serratus anterior 57
ulnar 58, 60
Nervous system 6, 16, 26, 37
Neurofibromata 18
Neurologic limps 66
Neurological examination 56, 88
Nutrition 13, 14, 22

Obesity 9, 14, 103
Occupation 7
Oedema 14, 15, 19, 22, 23, 103, 109
Opposition 49, 54, 59, 62
Orthopaedic surgical procedures 7

Osteoarthritis 3, 21, 22, 42, 52,
73–75, 81, 104, 105
Osteogenesis imperfecta 7, 9, 41, 68
Osteomalacia 95, 97, 99, 101
Osteomyelitis 13, 99–101
Osteopetrosis 92
Osteophytes 74, 104, 105
Osteoporosis 1, 7, 9, 10, 29, 96, 97,
104
Osteotomies 7

Paget's disease 5, 68, 95, 96, 101,
105, 107
Pain
aggravation 3
commencement 3, 4, 6
nature 3, 4
periodicity 3
radiation 3, 11
relief 3
severity 2
site 2
Painful arc syndrome 52
Pallor 14, 42
Palpitations 8
Paralysis 4, 5, 7, 16, 28, 30, 35, 56,
57
Parasitic agents 2
Parathormone 96
Parathyroid hormone 97
Parkinsonism 9, 16
Passive movements 50
Patella 64, 65, 68, 72, 73, 82
Patella alta 72
Patient
age 1
attitude 17
name 1
occupation 7
sex 1
Pellagra 14
Pelvic obliquity 27
Pelvic tilt 66, 78
Peripheral nervous system 16
Peripheral neuritis 14, 41
Peripheral vascular disease 15, 21,
43, 73
Peritonitus 13
Petechiae 42
Phleboliths 103

Phocomelia   41
Phosphate   97
Photon absorptiometry   106
Pigeon toes   76
Pilonidal sinuses   26
Pinch   5, 58, 59
Plain radiographs   103
Plasma factors   93
Pneumothorax   103
Poliomyelitis   4, 7, 16, 35
Position of comfort   23
Poverty   14
Power grip   58, 59, 63
Presenting complaints   2
Pressure sores   19
Profundus tendons   55, 56, 62
Pronation   35, 52, 53, 64
Prostatic carcinoma   95
Protein electrophoresis   97
Prothrombin   94
Prothrombin index   94
Prothrombin time   94
Psoriasis   8, 15
PTH   97
Pulmonary embolism   15, 94
Pulses   15, 43, 73, 109

Q-angle   65

Radicular pain   3
Radicular signs   30
Radio-isotopes   107
Radio-opacities   103
Rectal examination   30
Recurvatum   18, 64, 85
Reflex sympathetic dystrophy   22, 23
Reflexes   37
Reiter's syndrome   9
Renal calculi   9, 96
Renal colic   16
Renal failure   97, 99, 110
Renal tubular acidosis   97
Renal tumour   9
Respiratory system   8, 12, 15
Rheumatic fever   8, 15, 18, 99
Rheumatoid arthritis   5, 7, 8, 43, 48, 74, 100
Rheumatoid factor   100
Rib hump   29
Rickets   41, 68, 95, 99, 101, 105

Rigidity   35
Rupture quadriceps femoris   82
Rupture tendo Achilles   83

Sacral plexus   88
Sarcoidosis   97, 99
Scars   19, 20
Scheuermann's disease   28
Sciatica   3, 34
Scissored gait   67
Sclerae   18
Scoliosis   18, 27, 28, 29, 31
Scurvy   14
Senility   16, 42
Sensation   37
Sepsis   46
Septic arthritis   7–9, 22, 99
Septic shock   110
Septicaemia   13, 42, 95, 97, 110
Serum proteins   97
Sex linked   1
Shock lung   15, 42, 95, 110
Short leg limp   66
Shortness of breath   8
Shoulder girdle   46
Sickle cell disease   92
Single palmer crease   42
Sinography   106
Sinuses   20, 26
Sitting ability   27
Skin   8, 15, 18–21, 26, 64, 67, 109
Skin colour   15
Skin turgor   14
Slipped capital epiphysis   67
Social   1, 2, 6, 10
Softening   4, 5, 41, 76, 104, 105
Sonography   106
Spasm   3, 82, 85
Spastic contractures   67
Spasticity   4, 16
SPECT scan   107
Spinal deformities   28, 29, 31
Spinal instability   33
Spinal movements   31
Spinous processes   26
Splinter haemorrhages   42
Splints   3, 40
Spondylodysplasia   17
Sprain   2, 3, 66, 85
Sprengel's shoulder   46

Stance 65–67
Steatorrhoea 96, 97, 101
Steroid therapy 8
Steroids 9
Still's disease 8
Strokes 9, 48
Stunted growth 17
Subcutaneous fat 14, 40, 67
Sublimis tendons 55, 56, 62
Summary
    general orthopaedic examination 24
    examination of the upper limb 61
    examination of the lower limb 89
    examination of the spine 38
    orthopaedic history 10
Supination 35, 52, 53, 64
Surgical history 7
Swan-neck deformities 46
Swelling 6, 19, 22, 43, 72, 108
Syndactylism 41
Synovial fluid 22, 73
Synovial thickening 22, 41, 43, 74
Synovium 74
Syphilis 100

Tabes dorsalis 100
Tabetic gait 67
Tachycardia 15
Tachypnoea 15
Tarsal 72
Technitium 107
Telescoping 83
Temperature 20, 31, 43, 73
Temporomandibular joints 8
Tenderness 15, 21, 31, 72, 73, 90
Tendon ruptures 48
Tendon transfers 7
Texture 18, 26, 67
Thalassaemia 92
Thoracic spine 27, 28, 29, 31, 32
Thumb 54
Thumb in palm 48
Thyroid carcinoma 96
Tibial torsion 76, 77
Tomography 105
Tone 31, 35
Tongue 14, 15
Tophi 18
Tremor 42
Trendelenburg gait 66

Triggering 45
Trophic changes 41
Tuberculin 100
Tuberculosis 20, 22, 30, 41, 100, 101
Tumoral calcinosis 99, 103
Tumour 1, 5, 9, 16, 30, 31, 74, 96,
    97, 105, 107
Two point discrimination 37

Ulcerative colitis 7, 9
Ulcers 19, 20
Ulna 44, 46
Ulnar paradox 61
Unbalanced muscle action 4, 81
Upper limb 40
Upper limb deformities 45
Upper motor neurone 35
Urate 98
Uric acid 8, 19, 72, 98
Urinary aminoacid excretion 101
Urinary hydroxyproline 101
Urinary symptoms 9
Urinary tract infection 9

Valgus 24, 64, 65, 69, 75, 76
Varus 24, 64, 69, 75, 76
Vascular injury 41, 110
Vascular occlusion 2
Vascular thrombosis 43, 109
Venography 106
Villo-nodular synovitis 22, 74
Vitamin C 14, 98
Vitamin D 96–99
Vitamin deficiencies 14
Volar flexion 53
Volkmann's ischaemic contracture 4
Von Recklinghausen's disease 18

Waiter's tip position 57
Walking 66–68, 76, 82
Walking sticks 6
Wasting 41
Weakness 16, 30, 66, 67, 71
White cell count 93
Winging of the scapula 57

XDPs 94

Yersinnia 100